THE LAST WORD

NewScientist

Edited by Mick O'Hare
with illustrations by Spike Gerrell

Oxford New York
OXFORD UNIVERSITY PRESS

Oxford University Press, Great Clarendon Street, Oxford OX2 6DP

Oxford New York
Athens Auckland Bangkok Bogotá Buenos Aires Calcutta
Cape Town Chennai Dar es Salaam Delhi Florence Hong Kong Istanbul
Karachi Kuala Lumpur Madrid Melbourne Mexico City Mumbai
Nairobi Paris São Paulo Singapore Taipei Tokyo Toronto Warsaw

and associated companies in Berlin Ibadan

Oxford is a registered trade mark of Oxford University Press

First published 1998

British Library Cataloguing in Publication Data
Data available

Library of Congress Cataloging in Publication Data
Data available

ISBN 0-19-286199-9

3 5 7 9 10 8 6 4

Typeset in 9.5/12pt Meta Normal
by Graphicraft Typesetters Limited, Hong Kong
Printed in Great Britain by
Cox & Wyman,
Reading, Berkshire

CONTENTS

INTRODUCTION

The world is a mysterious place. There are big mysteries. How did the Universe begin? What is the nature of life? Who were our ancestors? All around the world, billions of pounds are spent trying to answer these questions. Huge space telescopes probe the furthest reaches of the Universe. Armies of laboratory scientists analyse the workings of our genes and try to unravel the enigma of life. Libraries of academic journals fill with the rapidly accumulating answers.

And then there are small mysteries. Why is the sky blue? Why do we all have different fingerprints? Why do men have nipples when they serve no useful purpose? When a tiny fly crashes head on with a speeding train, does it make the train stop for the tiniest fraction of a second? It is the small mysteries that are the subject of this book.

Small mysteries are everywhere. What is the purpose of that strange little bag on aircraft emergency oxygen lines? Why doesn't superglue stick to the inside of its tube? To those with a heightened sense of mystery, even absence can present a puzzle. How many times while walking a tree-lined street at night have you witnessed a sleeping bird plummeting from the branches above? 'Not once', will be the answer. So how is that birds can sleep soundly in the trees without ever losing their balance?

If you want to know the answer to this question and the others above, you can find them in the pages that follow. They are collected from the Last Word, the very last page of the weekly science magazine *New Scientist*. They are dedicated to everyone with a rich sense of curiosity and published with particular thanks to the thousands of readers who have sent in questions and answers to the Last Word over the past four years.

The Last Word page was officially launched in March 1994 after a *New Scientist* reader asked us why the LED lights on his microwave appeared to bounce up and down while he crunched away on his breakfast toast. But before that, readers had been sending in queries about unusual but commonplace phenomena for years. When we could, we answered them ourselves. Occasionally we published them in our letters section. The breakfast toast was the last straw. We decided it was time to put our readers into the kitchen and enlist their help with a special page devoted to explaining mysterious everyday occurrences.

Back then, we never guessed how the Last Word would take off. The page now sits comfortably at the head of reader polls of the magazine's most popular sections. The number of new questions and responses to old ones arriving each week would easily fill a weekly magazine by themselves. And, now that The Last Word appears on *New Scientist*'s Planet Science website, the number of responses by e-mail has overtaken the vast number we were already receiving by post and fax.

Many seemingly simple questions are actually very complex and do not have obvious answers. Thankfully, *New Scientist*'s readers have shown an extraordinary willingness to experiment for themselves in order to solve intractable problems. The strange behaviour of cream atop a glass of Tia Maria has been studied, tasting sessions have been held to see if tea made with reheated water tastes noticeably different, the curious behaviour of beer cans in buckets of iced water at barbecue parties analysed, and numerous photographs taken of clouds, rainbows, and patterns in frozen puddles.

Now that four years have passed since the launch of the Last Word, we decided to celebrate its continuing success and the efforts of our readers with the publication of this book, featuring many of our favourite questions and answers. We hope you will enjoy it and that it will inspire you to find a small mystery outside your door, or in your kitchen or perhaps even growing in your bathroom (but note we've already explained the nature of the peculiar black mould that spreads from damp corners).

If inspired, you'll find the readership of *New Scientist* is waiting to answer your questions, and to listen to your answers to their questions. So start writing. The Last Word page depends on your input, and it can be contacted at *New Scientist* Editorial, 1st Floor, Wardour Street, London W1V 4BN, or by e-mail at lastword@newscientist.com. Who knows, you might be the star of the second volume of this book.

And watch out for the sleeping birds that do not fall.

Mick O'Hare

PLANTS AND ANIMALS

BIRDZZZ

Q ***Why do birds never fall off their perches when sleeping. Do they, in fact, sleep?***
GRAEME FORBES
KILMARNOCK, AYRSHIRE

A Birds have a nifty tendon arrangement in their legs. The flexor tendon from the muscle in the thigh reaches down over the knee, continues down the leg, round the ankle, and then under the toes. This arrangement means that, at rest, the bird's body weight causes the bird to bend its knee and pull the tendon tight, so closing the claws.

Apparently this mechanism is so effective that dead birds have been found grasping their perches long after they have died.
ANNE BRUCE
GIRVAN, AYRSHIRE

A Yes, birds do sleep. Not only that, some do it standing on one leg. And even more surprising, may be hypnotized into sleep at will. My myna bird was.

If you wish to try it, bring your eyes close to the cage, and use the hypnotist's 'your eyes are getting heavier' principles (not spoken) on your own eyes. Act as if you are gradually falling asleep and the bird will follow you, finally holding one leg up under its belly, tucking its head under its wing, and falling into a deep sleep.

What's more, most pet bird owners know that all you need to do to make your pet fall asleep is to cover the cage with a blanket to simulate night.
DAVID LECKIE
HADDINGTON, EAST LOTHIAN

A Birds do sleep, usually in a series of short 'power naps'. Swifts are famous for sleeping on the wing.

Because most birds rely on vision, bedtime is usually at night, apart from nocturnal species, of course.

The sleeping habits of waders, however, are ruled by the tides rather than the Sun.

Some other species are easily fooled by artificial light. Brightly lit city areas can give songbirds insomnia. A floodlit racetrack near my home gives an all-night dawn effect on the horizon causing robins and blackbirds to sing continuously from 2 a.m. onwards. Unfortunately, I don't know whether it tires them out as much as it does me...

Andrew Scales
Dublin

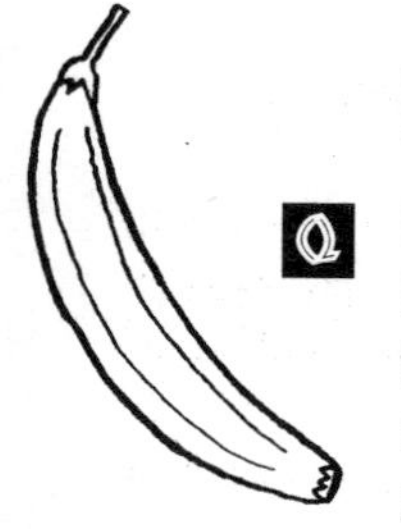

BANANA ARMOUR

The skin of bananas in the fridge turns brown faster than those in a room, but the fruit is still edible. I thought the browning was oxidation, but if so why does it happen faster in the cold?

Alun Walters
Cardiff

I wouldn't recommend putting bananas in the fridge to keep them fresh. Like all living organisms bananas adjust the composition of their cell membranes to give the right degree of membrane fluidity for the temperature at which they normally live. They do this by varying the amount of unsaturated fatty acids in the membrane lipids: the colder the banana, the greater the level of unsaturated fatty acid and the more fluid the membrane at a given temperature. If you chill the fruit too much, areas of the membrane simply become too viscous and the cell membranes lose their ability to keep the different cellular compartments separate. Enzymes and substrates which are normally kept apart therefore mix.

Overripe fruit kept out of the fridge goes brown by the same mechanism but, in this case, membrane breakdown occurs as a part of the general senescence of the tissue. In commercial storage chilling injury is, in fact, a big problem with tropical fruits, whereas temperate fruits like apples and pears can happily be stored at temperatures near freezing. I wonder therefore whether bananas stored in the fridge really taste as good as those left out? Incidentally, because tomatoes are a semi-tropical fruit I wouldn't suggest you keep them in the fridge either.

Alistair MacDougall
Institute of Food Research, Norwich

While many fruits are stabilized by refrigeration, most tropical and subtropical fruits (bananas in particular) exhibit chill injury. The ideal temperature for bananas is 13.3 °C . Below 10 °C spoilage is accelerated by the release of enzymes and the skin can blacken overnight as the banana fruit and skin soften. The enzymes leak from cellular storage sites and the leakage is caused by increased membrane permeability. This is mediated by ethylene gas which controls ripening and response to chill injury and such events as attack by parasites.

The two enzymes which break down the main polymers responsible for plant structure are cellulase and pectinesterase. These break down cellulose and pectin respectively. The breakdown of starch by amylase type enzymes is also involved in softening banana fruit tissue.

The blackening of the skin is caused by the release of another enzyme, polyphenyl oxidase

(PPO). This is an oxygen-dependent enzyme which polymerizes naturally occurring phenols in the banana skin into polyphenols similar in structure to melanin formed in suntanned human skin.

PPO is also inhibited by acid and this is why lemon juice is used to prevent browning in apples. Bananas are low in acidity and this may be one reason why they darken so quickly. Finally, blackening of the skin can be slowed by coating the banana with wax to exclude oxygen.

M. V. Wareing,
Braintree, Essex

Further to your previous answers: yes, the browning is an oxidation reaction. Yes, it is initiated by cooling. But no, the low temperature itself does not speed up the oxidation reaction in bananas.

Bananas like hot climates, and their cell membranes are damaged in the fridge. Membrane damage lets the phenolic amines such as dopamine, which are normally present inside the vacuoles of banana skin cells, leak out and encounter oxidizing enzymes (polyphenol oxidases) elsewhere in the cell. The dopamine can then be oxidized by atmospheric oxygen to form brown polymers, which may act as a defensive barrier. Once started by the cold-induced membrane damage, the browning reaction is promoted by warming.

For an extreme demonstration, put a banana skin in the freezer for a few hours. It stays creamy white because, although the membranes are shattered by freezing, oxidases cannot work at such low temperatures. Now let it thaw overnight at room temperature: the skin will go pitch black as the dopamine is oxidized. A control

banana skin kept at room temperature overnight stays whitish because the vacuolar membranes remain intact.
Stephen Fry
University of Edinburgh

TANK MADNESS

Q ***Following a recent bereavement we would like to know why fish jump out of small aquariums.***
Rowan White and Vicky
University of East Anglia

A Fish jumping out of small tanks is quite a common problem for enthusiasts, and is the reason why some owners choose to have a glass cover on the top of their aquarium. There are several theories as to why fish might jump from a small aquarium. It has been suggested that one reason fish leap from the water is that in the wild they use this method to attempt to rid themselves of ectoparasites.

Although the questioners did not mention the gender and species mix of the fish in their aquarium, it is possible that their fish could have been leaping to avoid predators or unpleasant interactions with other creatures, or even to show off to their conspecific fish, in some previously unknown courtship or territorial ritual. In the mean time, my sincere condolences for your loss.
R. Rosenberg
Stockholm, Sweden

A To captive fish, the air on the other side of the aquarium glass looks like water. And in

fish lore, the water is always cleaner on the other side.
John Chapman
North Perth, Australia

BAA-RMY

Q ***Why do sheep always run in a straight line in front of a car and not to the side?***
Aled Wynne-Jones
Cambridge

A Sheep and other animals run ahead of cars because they do not realize that cars cannot climb grassy banks. Ancestral sheep were pursued by wolves and big cats. If an animal tries to turn aside some yards from the hunter, the pursuing animal sees what is happening, makes an easy change of course, and intercepts the victim, which is presenting its vulnerable flank.

If, however, the prey dodges at the last minute, the outcome is different. The hare is the master of this strategy: as the greyhound reaches out with its jaws, the hare jinks to one side and the dog overshoots or, with luck, tumbles head over heels.

The instinctive response of a sheep or a hare to an approaching car is at least not as maladaptive as that of the hedgehog.
Christine Warman
Clitheroe, Lancashire

A Herbivores are killed by predators who normally grab them by the throat while running alongside, so the prey will always do its best to keep a potential threat behind its tail, swerving as the predator attempts to overtake. That's why a kangaroo, seeing a car drawing alongside, will jump onto the road right ahead in order to keep the car directly behind it, and often get run over in the process. As long as a car proceeds in a straight line behind a sheep, the sheep will try to outrun it in a straight line.

G. Carsaniga
Sydney, Australia

Sheep are much underrated. They don't merely run in a straight line—they run straight for a while, then dive to the side. This is not woolly thinking, it's perfectly logical. Sheep loose in the road are usually confined to country areas, where roads are bounded by steep verges, cliffs, hedges, fences, and ditches. The sheep recognizes that if it cannot beat the car on the flat, it stands no chance whatsoever up a bank, so it attempts to outrun the vehicle down the road.

What happens then is that the vehicle slows, and when it reaches a speed that is slow enough for the sheep to think it might beat the car over the obstructions at the side of the road, it swerves. And because most of the time this action is proved correct (most vehicles don't follow sheep off the road), the sheep carries on behaving in this way. QED, by sheep logic.

Clearly, this is a much more successful approach to road safety than that shown by humans. Humans rarely try to outpace the oncoming car. They tend to dive to the side of the road. Because more people are run over than sheep, one can conclude we have much to learn from sheep logic.

William Pope
Towcester, Northamptonshire

Sheep, being clever animals with an instinctive grasp of psychology, know that most drivers, though enjoying an occasional kill as long as they can use the 'it jumped in front of me, there was nothing I could do' excuse, are not so depraved as deliberately to run something down. Thus, running in a straight line has a distinct advantage over veering to the side.

Erik Decker
National Institute of Animal Husbandry
Department of Cattle and Sheep
Tjele, Denmark

FRIED FISH

Q ***My young neighbour asked me what happens when lightning strikes water. Do all the fish die? And what happens to the occupants of metal-hulled boats?***

Chris Cooper
Kempston, Bedfordshire

A When a bolt of electricity, such as a lightning bolt, hits a watery surface, the electricity can run to Earth in a myriad of directions.

Because of this, electricity is conducted away over a hemispherical shape which rapidly diffuses any frying power possessed by the original bolt. Obviously, if a fish was directly hit by lightning, or close to the impact spot, it could be killed or injured.

However, a bolt has a temperature of several thousand degrees and could easily vaporize the water surrounding the impact point. This would create a subsurface shock wave that could rearrange the anatomy of a fish or deafen human divers over a far wider range—tens of metres.

If someone in a metal-hulled boat was close enough to feel the first effect they would be

severely buffeted by the second. Besides which, metal hulls conduct electricity far better than water, so a lightning bolt would travel through the ship in preference to the water.
Andrew Healy
Ashford, Middlesex

A When lightning strikes, the best place to be is inside a conductor, such as a metal-hulled boat, or under the sea (assuming you are a fish).

Last century, the physicist Michael Faraday showed that there is no electric field within a conductor. He demonstrated this by climbing into a mesh cage and then striking artificial lightning all over it. Everybody except Faraday was surprised when he climbed out of the cage unhurt.
Eric Gillies
University of Glasgow

BLOWFISH

Q ***Fish don't fart, why is this?***
Christine Kaliwoski
Brentwood, California

A The writer probably thinks that fish don't fart because she has not seen a string of bubbles issuing from a fish's vent.

However, fish do develop gas in the gut, and this is expelled through the vent, much like that of most animals. The difference is in the packaging. Fish package their excreta into a thin gelatinous tube before disposal. This includes any gas that has formed or been carried through digestion. The net result is a faecal tube that either sinks or floats, but as many fish practise

coprophagia, these tubes tend not to hang around for too long.
Derek Smith
Long Sutton, Lincolnshire

A I have on several occasions witnessed my cichlids passing wind—to the displeasure of my plub eel.

This seems to be a result of them taking in too much air while wolfing down flaked foods floating on the surface of the water. If the air was not expelled it would seriously affect their balance.
Peter Henson
University of London

A Most sharks rely on the high density lipid squalene to provide them with buoyancy, but the sand tiger shark, *Eugomphodus taurus*, has mastered the technique of farting as an extra buoyancy device. The shark swims to the surface and gulps air, swallowing it into its stomach. It can then fart out the required amount of air to maintain its position at a certain depth.
Alexandra Osman
London

COLD FEET

Q ***Why do Antarctic penguins' feet not freeze in winter when they are in constant contact with the ice and snow? Years ago I heard on the radio that scientists had discovered that penguins had colateral circulation in their feet that prevented them from freezing but I have seen no further information or explanation of this. Despite asking scientists studying***

penguins about this, none could give an answer.

Susan Pate
Enoggera, Queensland

Penguins, like other birds that live in a cold climate, have adaptations to avoid losing too much heat and to preserve a central body temperature of about 40 °c. The feet pose particular problems because they cannot be covered with insulation in the form of feathers or blubber, yet have a big surface area (similar considerations apply to cold-climate mammals such as polar bears).

Two mechanisms are at work. First, the penguin can control the rate of blood flow to the feet by varying the diameter of arterial vessels supplying the blood. In cold conditions the flow is reduced, when it is warm the flow increases. Humans can do this too, which is why our hands and feet become white when we are cold and pink when warm. Control is very sophisticated and involves the hypothalamus and various nervous and hormonal systems.

However, penguins also have 'countercurrent heat exchangers' at the top of the legs. Arteries supplying warm blood to the feet break up into many small vessels that are closely allied to similar numbers of venous vessels bringing cold blood back from the feet. Heat flows from the warm blood to the cold blood, so little of it is carried down to the feet.

In the winter, penguin feet are held a degree or two above freezing—to minimize heat loss, while avoiding frostbite. Ducks and geese have similar arrangements in their feet, but if they are held indoors for weeks in warm conditions, and then released onto snow and ice, their feet may freeze to the ground, because their physiology has adapted to the warmth and this causes the blood flow to the feet to be virtually cut off and their foot temperature falls below freezing.

John Davenport
University Marine Biological Station
Millport, Isle of Cumbrae

I cannot comment on the presence or absence of colateral circulation, but part of the answer to the penguin's cold feet problem has an intriguing biochemical explanation. The binding of oxygen to haemoglobin is normally a strongly exothermic reaction: an amount of heat (DH) is released when a haemoglobin molecule attaches itself to oxygen. Usually the same amount of heat is absorbed in the reverse reaction, when the oxygen is released by the haemoglobin. However, as oxygenation and deoxygenation occur in different parts of the organism, changes in the molecular environment (acidity, for example) can result in overall heat loss or gain in this process.

The actual value of DH varies from species to species. In Antarctic penguins things are arranged so that in the cold peripheral tissues, including the feet, DH is much smaller than in humans. This has two beneficial effects. First, less heat is absorbed by the birds' haemoglobin when it is deoxygenated, so the feet have less chance of freezing.

The second advantage is a consequence of the laws of thermodynamics. In any reversible reaction, including the absorption and release of oxygen by haemoglobin, a low temperature encourages the reaction in the exothermic direction, and discourages it in the opposite direction. So at low temperatures, oxygen is absorbed more strongly by most species' haemoglobin, and released less easily. Having a relatively modest DH means that in cold tissue the oxygen affinity of haemoglobin does not become so high that the oxygen cannot dissociate from it.

This variation in DH between species has other intriguing consequences. In some Antarctic fish, heat is actually released when oxygen is removed. This is taken to an extreme in the tuna, which releases so much heat when oxygen separates from haemoglobin that it can keep its body temperature up to 17 °C above that of its environment. Not so cold-blooded after all!

The reverse happens in animals that need to reduce heat due to an overactive metabolism. The migratory waterhen has a much larger DH of haemoglobin oxygenation than the humble pigeon. Thus, the waterhen can fly for longer distances without overheating.

Finally, fetuses need to lose heat somehow, and their only connection with the outside world is their mother's blood supply. A decreased DH of oxygenation by the fetal haemoglobin when compared to maternal haemoglobin results in more heat being absorbed when oxygen leaves the mother's blood than is released when oxygen binds to fetal haemoglobin. Thus, heat is transferred into the maternal blood supply and is carried away from the fetus.

Chris Cooper and Mike Wilson
University of Essex

FLYING FINS

Q ***Why do flying fish fly? Is it to escape predators, or to catch flying insects, or as a more efficient means of getting around***

than swimming? Is there some other entirely different reason?

Julyan Cartwright
Palma de Mallorca, Spain

The usual explanation for flight in flying fish is as a way to escape predation, particularly from fast-swimming dolphinfish. They do not fly to catch insects; flying fish are largely oceanic and flying insects are rare over the open sea.

It has been suggested that their flights (which are actually glides because flying fish do not flap their 'wings') are energy-saving but this is very unlikely as the vigorous takeoffs are produced by white, anaerobic muscle beating the tail at a rate of 50 to 70 beats per second, and this must be very expensive in terms of energy use. Flying fish have corneas with flat facets, so they can see in both air and water. There is some evidence to suggest that they can choose landing sites. This might allow them to fly from food-poor to food-rich areas, but convincing evidence of this is lacking. There seems to be little doubt that escape from predators is the major purpose of their flight, and this is why so many fly away from ships and boats, which they perceive to be threatening.

John Davenport
University Marine Biological Station, Millport, Strathclyde

Strictly speaking, the flying fish does not fly, it indulges in a form of powered gliding, using its tail fins to propel it clear of the water. It sustains its leap with high-speed flapping of its oversized pectoral fins for distances of up to 100 metres. The sole purpose of this activity seems to be to escape predators. If one can manage to tear one's eyes away from the magic of the unexpected and iridescent appearance of a flying fish, a somewhat

more substantial fish can often be seen following its flight path just below the surface.
Tim Hart
La Gomera, Canary Islands

A I have seen whole schools of flying fish become airborne as they try to escape tuna which are hunting them, and minutes later have seen the school of tuna attempt similar aerobatics as dolphins move in for their supper of tuna steaks.

A morning stroll around the decks of an ocean-going yacht will often provide a frying-pan full of flying fish for breakfast. Presumably they are instinctively trying to leap over a predator (in this case the boat) but as they don't seem to be able to see too well at night they land on the deck. They rarely land on board during the day. Most alarmingly they will land in the cockpit, and even hit the stargazing helmsman on the side of the head.
Don Smith
Cambridge

Breaking the mould

Q ***What is the name of the dreaded black mould that colonizes damp places in bathrooms? Because materials produced to remove the mould do not seem to work, and nor do household bleaches, detergents and solvents, can anyone suggest a remedy other than abrasives?***
G. W. Green
Malvern, Worcestershire

A The infamous black mould is the fungus *Aspergillus niger*. The reason it seems so difficult to eradicate is that the visible black manifestation is merely the exposed structure of the fungus, which is mainly comprised of the fruiting bodies. In addition to this visible material, there is invariably an insidious network of hyphae or mycelia which lie in the substrate of wallpaper or plaster, and feed on the minerals contained within.

Eradication of the mould requires not only the repeated physical removal of the visible growth but the simultaneous use of a penetrative fungicide capable of permeating the substrate and killing off the unseen root structure. The analogy is that of trying to eradicate ground elder or horsetail from your vegetable patch by merely strimming the visible plantlets.

Andrew Philpotts
Hexham, Northumberland

The *Aspergillus* fungus has been a constant source of annoyance in local council accommodation throughout the nation. It is prevalent where cool, still air deposits condensation next to steel window frames, concrete-screed ceilings, water-tank enclosures, and similar areas.

Current medical opinion is that this fungus is a major source of allergenic disease and that it produces carcinogenic aerosols, so removal of the unsightly nuisance is also important for health.

I have experienced problems when attempting to remove *Aspergillus*. Table salt and bleach have only limited success, but I finally effected a permanent solution by washing the affected areas several times with a systemic fungicide available from any garden store. This may not, however, be the safest solution as the fungicide may be as toxic as the fungus.

Glyn Davies
Kingston, Surrey

The previous answers seem to have fallen into the trap of assuming that any mould that is black is

Aspergillus niger. In surveys of Scottish housing carried out by my laboratory, the incidence of this species has been rather low.

The most common dark mould in growths on bathroom and other damp walls are likely to be species of *Cladosporium*, with *Aureobasidium*, *Phoma*, and *Ulocladium* thrown in for good measure. Even green species of *Aspergillus* and *Penicillium can* look black when soaked.

The situation is much the same in mainland Europe, so it is likely that it will not be very different in Northumberland or Surrey, unless the bathrooms there come closer than most British bathrooms to providing the subtropical and tropical climates that favour *A. niger.*

A really black fungus in about 15 per cent of houses in Scotland with mould problems is *Stachybotrys aira* (syn. *S. chariarum*). Wallpaper, jute carpet backing, or the cardboard wrapper of gypsum board all provide ideal cellulosic substrates on which it can thrive in very damp conditions. This type of mould may present the greatest hazard of any to the health of occupants of mouldy buildings. Its airborne spores are allergenic and powerfully toxicogenic. Its toxins inhibit protein synthesis, and are immunosuppressive, an irritant, and haemorrhagic.

It is well known that fodder contaminated by *Stachybotrys* can kill horses, and it is also harmful to the stable hands. Currently, this mould is of particular concern in North America, where it has been implicated in episodes of building-related illness ranging from chronic fatigue syndrome in adults to fatal pulmonary haemosiderosis in infants. Consequently it has been the subject of lawsuits (one for the sum of $40 million) taken out against builders and employers.

Brian Flannigan
Department of Biological Sciences, Heriot-Watt University, Edinburgh

A The outside wall of my bathroom in Pimlico used to grow a superb crop of mould which removed the wallpaper and infested the plaster. To remove this *Aspergillus*, I used a single washing down with a dark pink solution of potassium permanganate crystals which effected a cure with no recurrence.

Bill Christie
Fairlight Cove, East Sussex

Readers should note that potassium permanganate is poisonous if swallowed—Ed.

A Household bleach does not remove the staining caused by *Aspergillus niger*. But spraying the surface or painting it with an aqueous solution of 10 per cent zinc sulphate prevents the re-emergence of the fungus as long as the molecules of zinc sulphate are not washed off.

Farrokh Hassib
London

USE YOUR NUT

Q ***How does a squirrel find the peanuts that it has buried in the ground?***

Xiang Li
Liverpool

A The American animal psychologist, Edward Tolman, showed that animals form a cognitive map of the environment that they are exploring. In other words, information about features, landmarks, and spatial relationships of the environment is stored in their brain. Observation of their feeding strategies suggests that they also form representations of the type of food to be found in certain sites. For example, the nutcracker bird can remember locations at which it has hidden food for a long time afterwards and squirrels use the same methods.

Joanna Seed
Liverpool

It would seem that because mammals have a more highly developed olfactory sense than that of food-storing birds, mammals that scatter-hoard their food (that is, distribute many food items over a wide space) would depend on smell to find their caches.

Fox squirrels have been shown in field tests to be able to find artificially made caches quite readily and visual cues have frequently been shown to be unimportant in cache recovery by rodents. However, in some situations, grey squirrels seem to rely on spatial memory to retrieve nuts. When they were allowed to cache hazelnuts in an outdoor arena, they found their own caches more often than would be allowed by pure chance, but they could not find control hoards placed by the researcher.

And red squirrels have been observed to recover with some accuracy stored fungi from pine trees. The stores could not be seen by the squirrels and it seems unlikely that olfactory cues were important because the fungi were stored high up in the trees.

This all suggests that spatial memory may be relied on by squirrels, but more experiments are needed to determine to what extent and whether this is the case with other food-storing rodents.

Anthony McGregor
University of Newcastle

POLAR BEAR

Q ***Why are there so many types of animals in the jungle and so few at the poles?***
5-YEAR-OLD BOY
NO ADDRESS SUPPLIED

A A general principle of ecology is that harsher environments provide fewer suitable niches for plant and animal species. This is clearly seen as one moves from a lower elevation to a higher one in the mountains of the north-western USA.

Plant lists made at lower elevations may contain a hundred or more species, while at or near the tree line the list may dwindle to just a dozen or under. Fewer plants equate to fewer ecological niches and fewer species of animals to fill them.
RICHARD BARTH
WALLACE, IDAHO

A There can only be a small number of animals there because there is little sunlight, few plants, and not much food. If instead of the polar bears there were 57 different varieties of animal, there would be so few of each that they would have difficulty finding mates, and the smallest run of bad luck would wipe a species out.
TIM STEVENSON
BRACKNELL, BERKSHIRE

A At the poles, there is a very small number of environments for animals to live in—on snow, in snow, on bare rock, at the coastline, on ice, and in the ocean.

Jungles, with their many layers of plant life, provide an almost limitless number of habitats for animals. With the wide diversity of available food and living accommodation, diversity among the young is promoted.

Bill Woodelkl
No address supplied

NOT MUSH ROOM

Q ***Near where I live there are toadstools growing through the pavement, the surface of which they have displaced in fairly large chunks. What mechanism allows toadstools—essentially very soft and squashy items—to push through two inches of asphalt?***

John Franklin
London

A The toadstools forcing their way up through asphalt are probably ink-cap mushrooms (species *Coprinus*) growing on buried plant debris. They are pushing upwards because their stalks function as vertical hydraulic rams.

The upwards pressure comes from the turgor pressure of the individual cells making up the wall of the hollow stalk of the mushroom. Each individual cell grows as a vertical column by inserting new cell wall material uniformly along its length.

The major structural component of the cells is a shallow helical arrangement of fibres of chitin winding round the axis of the cell. These chitin fibres are embedded in matrix materials, making the wall material like a carbon fibre composite. Chitin is an exceptionally strong biopolymer (also used by insects for their exoskeletons), and gives immense lateral strength to the fungal cell wall, so that internal pressure is confined as a vertical column. Water enters the cell by osmosis, and the resulting turgor pressure provides the vertical force that allows the mushroom to push up through the asphalt.

This phenomenon was first investigated 65 years ago by Reginald Buller, who measured the lifting power by loading weights on to a mushroom that was elongating inside a glass tube. He calculated an upwards pressure of about two-thirds of an atmosphere.

The cells have a gravity-sensing mechanism that keep the mushroom exactly vertical. A mushroom that is put on its side will rapidly reorient to grow vertically again.

Graham Gooday
University of Aberdeen

Two inches of asphalt is nothing to the muscular mushroom. One large shaggy ink-cap (*Coprinus comatus*) discovered at Basingstoke lifted a 75 by 60 centimetre paving-stone 4 centimetres above the level of the pavement in about 48 hours.

Historically, mushrooms often sprang up in foundries, supposedly from horse manure used in preparing loam for casting, and were often reported as having lifted heavy iron castings. Presumably these would have been some type of field mushroom such as *Agaricus campestris*. Whatever the species, the mechanism by which the force was exerted is likely to be the same, namely hydraulic pressure.

As Buller found, the exquisite and fragile *Coprinus sterquilinus* exerts an upward pressure of nearly 250 grams with a stem 5 millimetres

thick, so it is not surprising that more robust species can tear the tarmac.
RICHARD SCRASE
MUSHROOM MAKERS
BATH

FISHING LIMITS

Q ***What is the limiting factor for the size of a fish? Does the same factor apply to mammals such as whales? If so, why is it that the largest known fish is smaller than the largest whale?***
DUNCAN FISHER
BRENTWOOD, ESSEX

A The most likely reason why whales are larger than the largest fish is the way they take up oxygen to fuel their metabolism. Fish have to absorb oxygen from water via their gills while whales absorb air via their lungs. As M. Jobling points out in *Dissolved Oxygen, Fish Bioenergetics* (1994), chapter 15, p. 231, seawater at 10 °c with an oxygen saturation of 50 per cent contains approximately 3.97 millilitres of oxygen per litre, which is equivalent to an oxygen tension of 78.9 millimetres of mercury. At higher water temperatures this oxygen tension will be even lower (think about fish that try to gulp air from the water surface on a hot summer day). The air immediately above sea level would have an atmospheric pressure of 760 millimetres, which at an oxygen concentration of 20.9 per cent translates to 159 millimetres of oxygen pressure. Therefore, in seawater at 10 °c you would expect around 4 millilitres of oxygen per litre, but certainly not more than 10 millilitres compared to 209 millilitres found in air. Therefore, the whale has access to 20 to 50 times more oxygen when breathing in air than a fish can absorb through its gills. Oxygen is thus likely to become a limiting factor as fish grow, even with gills, the most efficient mechanisms designed in nature to absorb oxygen from water.

M. Pannevis
Verden-Aller, Germany

One possibility is that the fish is limited by its heart whereas the mammal is not. Fish hearts do not have a left and right side as mammals' hearts do. The right side of the heart in mammals pushes the blood at low pressure through the lungs (so the thin walls between the capillaries and lungs do not burst). The blood then enters the left side and is pumped at high pressure through the whole body.

In fish the blood flows directly from the gills to the body, so the pressure must be low enough not to burst the blood vessels in the gills but still high enough to push blood through the whole body. The larger the body the more pressure is required.

Peter Lange
No address supplied

MYSTERIES AND ILLUSIONS

POLE POLL

Q ***What time is it at the North Pole?***
Nigel Goodwin
Nottingham

A There are two answers to this question. The first is that the time for a person is the time determined by her or his circadian rhythm. Initially, this physiological time will be close to the time for the longitude where the individual lived before visiting the pole. Over a period of weeks at the pole, this time will drift as the individual settles into a rhythm with a period that is usually about 25 hours long.

Of course, there is also a local time, independent of people, unless you are a philosopher residing somewhere other than the pole.

So the second answer is that the time is either daytime (for the six months of summer) or night-time (for the six months of winter).

I have not been at the pole near the equinoxes, but I would imagine that there are also several continuous weeks of twilight when the sun is just below the horizon.
Will Hopkins
University of Otago, New Zealand

A The crux of the question is this: how should a person born and bred at the North Pole, who has never heard of Greenwich or Tokyo or any other place on the Earth, start measuring time?

This can be done as follows. Suppose that it is the dark period of the North Pole when the Sun is below the horizon all the time. Fix a horizontal board at the pole and draw a circle on it with two diameters perpendicular to each other. Label the ends of the diameters A, B, C, D round the circle.

At the North Pole, you can see the stars revolving in planes that are parallel to the horizon. The plane of the horizon coincides at the poles with that of the celestial equator.

Then choose some star on the horizon and define zero hour as the instant when this star passes across the line of sight of point A when viewed from the circle's centre (the pole). The crossings of the star across B, C, and D will correspond to 6, 12, and 18 o'clock respectively.

It is then easy to draw other straight lines on the board representing intermediate hours.

If I were required to do this exercise at present (at the North Pole), for reference I would choose the faintest of the three stars in Orion's belt because it lies almost exactly on the celestial equator, is the brightest of all the stars that lie that near, or on the celestial equator, and it is clearly visible to the naked eye.

The next problem at the Pole would be how to decide what time it was in the summer, when no stars can be seen because it is constantly daylight.

Having drawn the hour lines in the winter, you must wait for the Sun to appear above the horizon. At the moment it is sighted at the approach of the Arctic spring, make a note of its direction on the board. The hour line on which it falls can be called the time of sunrise in the 24 hours system that was devised during the winter.

The Sun will revolve, like the stars in winter, in a plane parallel to the horizon, but unlike our reference star, which always revolves in the same plane, the Sun's plane will be higher up day by day, ultimately reaching a highest level of 23.5 degrees from the horizon.

Then it will become lower and lower again until, six months after our first sighting, the Sun will disappear below the horizon.

D. S. Paransis
Luleå University of Technology, Sweden

This isn't a sensible question: time is independent of location. When it is 1800 GMT in London, it is also 1800 GMT at the North Pole, in Timbuktu, or on the far side of the Moon.

One could ask: 'what time zone is it at the North Pole?' but this also fails. Time zones are defined politically and administratively rather than by geography. Because the North Pole is floating on the high seas no time zone is defined for it.

Attempts to define a natural time astronomically also fail. Noon is when the Sun is due south, but at the North Pole the Sun is always due south. Noon is when the Sun is at its highest, but the height of the Sun is essentially constant at the North Pole. Noon is halfway through the period of daylight, but at the North Pole it is light for six months and then dark for six months.

Mike Guy
Cambridge

Time, from a geophysical point of view, is related to the position of the Sun over the Earth and to the position of the observer. Because any direction from the North Pole is south, the Sun is always in the south and whatever the time is at the North Pole, it is always the same time.

What time is that? The International Date Line runs through the North Pole, leaving the pole sitting eternally between one date and the next. In other words, it is always midnight at the North Pole.

This, of course, explains how Father Christmas manages to deliver presents to every good little

boy and girl throughout the world in the space of a single night. He just heads out of his grotto due south (which from the North Pole is any direction) drops off as many presents as he can fit on his sleigh and then he heads back home where it is exactly the same time as when he left. So he can then drop off more prezzies, return home, and so forth.
Patrick Whittaker
Hounslow, Middlesex

A The North Pole is, of course, the true spiritual home of the politician because in answer to the question 'what time is it?', she or he can, with all honesty, say 'what time do you want it to be?'
Paul Birchall
Mickelover, Derbyshire

IS IT A BIRD?

Q ***My patio is lit by the summer Sun, which illuminates the floor and adjacent walls. Last year I noticed that when a bird flew overhead, it cast a shadow on the patio. The bird was flying at the correct height to just clear the wall on my right. Its shadow moved across the patio floor at the same speed as the bird. However, when the shadow reached the base of the wall, it accelerated instantly, climbing the wall before reaching the top and continuing along the flat roof, again at the same speed as the bird. If the bird had been an object travelling at the speed of light, then the shadow on the ground and the flat roof would have also been travelling at the speed***

of light. But what of the period when the shadow was travelling up the wall? Under this condition, the shadow would have had to travel faster than light, which would present a direct challenge to Einstein. Is there any other explanation?

A. R. Seabrook
Eastbourne, East Sussex

A Yes, the shadow does indeed move faster than light. No, this is not a violation of the special theory of relativity, because you cannot transmit information using a shadow. This is one of the key points of the special theory. Information can be transmitted by light, but a shadow marks the absence of any light so no information is transmitted faster than light.

Many other examples can be described to show geometric objects moving faster than light. For instance, take two straight, crossed lines with an extremely small angle between them and move the end of one line slowly towards the other. The crossing point will move with increasing speed. If the angle is small enough this speed will be faster than the speed of light. However, again no information is transmitted and no material object which could be used to transmit information travels faster than light.

Alex Niemeyer
Munich, Germany

A The question does seem to indicate the possibility of movement faster than light (FTL). However, the illusion is generated by our viewpoint—nothing actually moves faster than the speed of light (c).

We can observe a similar effect in the 'lighthouse' paradox. Imagine the beam of a lighthouse rotating at 1 revolution per minute.

As distance from the lighthouse increases, the track of the beam, when observed striking a vertical surface, would seem to be moving faster as it passed along that surface. Calculations indicate that at a distance greater than approximately 3 million kilometres the beam would be rotating at FTL speed. Of course, the beam's track may be rotating thus, but the quanta making up the actual radiated light energy moving away from the lighthouse continue to travel at c.

To return to the bird, if the creature were to fly over the patio at c, all that would happen is that the roof would have the sunlight cut off from its surface in a time slightly less than it takes for light to reach the patio, similar to walls of increasing distances reflecting the lighthouse beam. This is the time taken for the light to travel the height of the wall.

John Smith
Leicester

A A bird ultra-speedy in flight
Won't give Doc Einstein a fright
Its shadow's an absence
Of light, not a presence
And nothing is faster than light

Doug Cross
Honiton, Devon

LEFT IN DOUBT

Q ***As a left-handed person I was both amused and annoyed by* New Scientist*'s article 'Sudden death for left-handers' which suggested that left-handed people are at greater risk of accidental death. How can this be? Surely a right-handed person has just as***

much chance of dying accidentally as I do. Or is there some unknown factor involved?
Alan Parker
London

A When approaching obstacles, right-handed people will, in general, circumvent them by going to the right, while left-handed people will go to the left. If two same-handed people approach an obstacle from the opposite direction they will walk safely around it without bumping into one another *en route*. If two people of different handedness approach an obstacle from the opposite direction, they will pass on the same side leading, potentially, to a bump. Because most people are right-handed, it is left-handed people who most frequently find themselves bumped in these situations. This is a simple example, but taken to extreme and multiplied by a lifetime of bumps, the result is a shorter life expectancy for left-handed people.
Hannah Ben-Zvi
New York

A We left-handers are at greater risk of accidental death because industrial tools and machinery are designed for the right-handed. Left-handers are, therefore, more likely to chop off parts of themselves in all manner of mechanical devices.

An interesting example is the SA-80 assault rifle. When fired from the left shoulder it ejects spent cartridges, at great velocity, into the user's right eye.
Daniel Bristow
Kew, Surrey

Aisle Miles

Q ***Two people lose each other while wandering through the aisles of a large supermarket.***

The height of the shelves precludes aisle-to-aisle visibility. One person wishes to find the other. Should that person stop moving and remain in a single visible site while the other person continues to move through the aisles? Or would an encounter or sighting occur sooner if both were moving through the aisles?

David Kafkewitz
Newark, New Jersey

A The best strategy may be to wait at the exit of the store on the grounds that the other person may eventually conclude that you have gone home and do likewise. The maximum waiting time will then be from the time you lost each other until the store closes.

A strategy of staying still only works if just one person stays still. If you both decide to stay still, then the wait time is either infinite, if you get locked in, or again until the store closes.

Assuming that one person stays still while the other searches, then the maximum time is the time taken for one person to search the entire store. This depends on the layout of the store: if all the aisles can be readily seen from one vantage-point, then the search is simplified. The problem is not dissimilar to that of designing prisons in which the warders can see down as many corridors as possible, or the design of forts that will give the defenders maximum cover. In order to increase the odds of being located, the person staying still should stand at an intersection of aisles.

A random search will proceed with each person moving away from their initial starting-point at a rate proportional to the square root of time. The area being searched by each person is defined by two circles centred at their respective search starting-points. Given that these circles will need to overlap significantly for the individuals to meet, the search time must be at least proportional to the square of their initial separation distance. If some of the aisles are blocked during this search then the rate of movement is reduced and the problem becomes one of motion on a fractal where the rate is proportional to some fractional exponent.

Stephen Massey
St Albans, Hertfordshire

To begin to answer this question, one must first know whether the two people have agreed in advance what to do if separated—for example, who should wait and who should search. If they can agree on independent search strategies in advance, the problem is the asymmetric version of the rendezvous search problem (see below); otherwise it is the symmetric version.

I discuss both versions of the problem in a paper to be published in the Society for Industrial and Applied Mathematics *Journal of Control and Optimization*, and several specific cases in particular search regions have subsequently been solved. In all these cases where exact solutions (to give a least expected time or minimax time) have been obtained, both searchers move at their maximum speed all the time. In these cases it is certainly not optimal for a searcher to stop while the other continues. For example, in a simplified model in which two people are placed a unit distance apart, but neither knows the direction of the other, it would take an expected time of $(1 + 3)/2 = 2$ for the searcher to find the stationary person (assuming visibility is nil). However, by moving optimally this time can be reduced to 13/8.

The only case I know of where a searcher and a waiter may be optimal is for two people placed randomly on a circle, and then only when the people concerned have no common notion of clockwise; otherwise one person walking clockwise and the other anticlockwise is optimal.

All these results and questions assume that the searchers find each other only when they meet or alternatively when they come within a specified detection radius. This applies to aisles in a supermarket on a crowded day, when visibility along an aisle is limited. The possibility of seeing a long distance along an aisle has not, as far as I know, been modelled.

In case anyone is interested, the full bibliography on this topic is: 'The rendezvous search problem', S. Alpern; 'Rendezvous search on the line with distinguishable players', S. Alpern and S. Gal; 'Rendezvous search on the line with indistinguishable players', E. Anderson and E. Essegaier. All three papers appear in the SIAM *Journal of Control and Optimization* (1995).

Steve Alpern
London School of Economics

I recommend that you walk along the edge of the supermarket where the tills are, looking down the aisles for the person you seek. If you have no success, then walk back, still looking down the aisles, but also checking the tills. If you still have no success, then find the cold meat counter, as queues often develop there. Then have a final walk along the till edge, checking the aisles again. If you are still unsuccessful then you should ask

for an announcement to be made on the public address system—or, if it's not urgent, wait by the exit.
Owen Crossby
Aberystwyth, Dyfed

A I suggest you head towards the security camera and use it to track the other shopper.
Joyce Sumner
Oldham, Lancashire

A Failing any of the above options, you must consider the alternatives: (*a*) the other person has taken their chance to escape from you, (*b*) they have been removed for shoplifting or worse, (*c*) they are playing games with you, such as walking behind you. It helps enormously to have brought up children. The invisible intuitive antennae developed as a result of prolonged exposure to children seem to last for the rest of your life.
Sophie Hyde Parker
Taunton, Somerset

SPLAT

Q ***The following paradox has puzzled me since I was a child. A fly is flying in the opposite direction to a moving train. The fly hits the train head-on. As the fly strikes the front of the train, its direction of movement changes through 180°, because it hits the windscreen and continues as an amorphous blob of fly-goo on the front of the train.***

At the instant it changes direction, the fly must be stationary and because, at that

instant, it is also stuck on to the front of the train, the train must also be stationary. Thus a fly can stop a train. Where is the logical inconsistency in this (or does it explain something about British Rail)?

Geoff Fleet
Evanston, Illinois

You are right. A fly does stop a train, but not the whole train, just part of the small local area where the fly makes contact, and then not for very long.

All objects, no matter how rigid they seem, are flexible to some extent. So the train's windscreen, on being struck by the fly, deflects backwards very slightly. That small piece of train not only stops for a short period but can actually move backwards. It takes considerable force to do this (glass being fairly rigid) but it should be remembered that the forces involved in any type of impact are typically quite large.

The force exerted by the fly on the train is the same size as the force exerted by the train on the fly—a large force. And such a force acting on the small mass of the fly gives rise to a very large rate of acceleration. In fact, the rate of acceleration of the fly is so great that it accelerates up to the speed of the train in only the short distance by which the windscreen has been deflected.

Having got the fly up to speed, the windscreen then springs back to its original shape. Because it springs back very quickly the deformed part actually overshoots its original position and a vibration is then set up as it springs back and forth trying to regain its original form. This gives rise to the sound we hear when the fly hits the windscreen.

This simple picture is complicated by factors such as the crushing of the fly's body and inertia

effects in the glass, but it does demonstrate the principles that are involved.
Eric Davies
Perth, Australia

The questioner is correct in the assumption that the fly must, at some point, be stationary. However, at this point it is not 'stuck' onto the front of the train. As soon as the train's front window touches the front of the fly (ignoring the effect of the wall of air that is pushed in front of the train), the fly is accelerated forward in relation to the train. During the very short, but finite, period of time that it takes the train to cover the length of the fly's body, the fly is being compressed and accelerated.

Thus, in the instant that the fly is stationary, perhaps its front 10 per cent has become goo on the train's window. The train has maintained a constant speed during this process. By the time the front of the train has completely caught up with the whole of the fly, some 2×10^{-4} seconds later at 200 kilometres per hour, the fly has been accelerated up to the speed of the train and continues, now completely flattened, to move with it.

A slightly more pedantic point is that, by conservation of momentum, the train will be very slightly slowed, although it will quickly build up to its original speed. The acceleration felt by the fly, if accelerated by 200 k.p.h. over 1 centimetre, is around 3×10^5 metres per second per second—about 30 000 g. The force felt by a 1 gram fly and the window is around 300 newtons.
Julian Bean
Richmond, Surrey

When the train hits the fly, the front few nanometres of the windscreen's impacting surface stop momentarily, the next few nanometres suffer elastic deformation, and the rest of the train continues at full speed.

After the impact, the compressed windscreen material will recover, accelerating its front edge up to the full speed again, and showing virtually no sign of damage (unlike the inelastic deformation of the fly).

This is a slight oversimplification as, in practice, an elastic stress wave will propagate backwards into the train, and the front surface will oscillate until the motion is cancelled out, but such effects will be unimportant in the case of the fly and the train. Where the masses are more equal, as in the case of colliding cars, the additional motions within each structure may be important as, for example, they may determine the type of injuries suffered by the occupants.

M. G. Langdon
Farnham, Surrey

Readers' explanations about the fly hitting the train cover many aspects from the width of the fly to the pliability of the windscreen. (What if the fly hits the boiler instead?)

But they completely miss the implied point of the question, which is philosophical rather than physical. For 'fly' substitute 'one atom of the fly'. This is just a rerun of the paradox posed by Zeno of Elea. Around 450 BC he said that a moving object is always in motion, and yet at any given time it is somewhere (that is, stationary). We humans cannot see, measure, or imagine an infinitely small time any more than we can truly imagine infinity. We never will.

R. K. Hendra
London

THE DATING GAME

If I was transported in a time machine to an unknown date in Australian history that preceded any European contact on that continent, what would be the best way to determine the date?

Douglas Nichols
South Hobart, Australia

I'd suggest the following. Take a good look at the Moon. Are the familiar bright ray craters—Copernicus and Tycho—missing? Too bad—you've been transported more than one hundred million years into the past, and determining the date is practically impossible.

Is the Moon OK? Great—you're not as far back in time. Next, take a look at the constellations. Do they look unfamiliar? If so, you may have travelled hundreds of thousands of years into the past, long enough for the proper motions of stars to distort the shapes of the constellations. You'll still have no luck determining the date.

If the constellations look right, try to determine the location of the south celestial pole between the stars, and compare this with its twentieth-century location. Because of the precession of the rotational axis of the Earth, the celestial pole traces a complete circle among the stars every 25 700 years. Observations will enable you to find the millennium you are in.

Next, find out where the bright planets Jupiter and Saturn are, relative to each other, relative to the Sun, and relative to the constellations of the zodiac. Jupiter traverses the 12 constellations of the zodiac in 11.86 years; Saturn in 29.46 years.

This means that their relative position with respect to the zodiac and to the Sun repeats once every 1357 years (114 Jupiter revolutions equal 1351.84 years; 46 Saturn revolutions equal 1355.16 years); if you know the planets' position in, say, 1995, and you know the millennium you're in, you can work out the year.

If you happen to be carrying your laptop computer with you, load your favourite astronomical software and check your results by looking at the positions of the other naked-eye planets (Mercury, Venus, and Mars) or by watching a lunar or solar eclipse.

Finally, if you've got the year right, finding the exact date is simple. Determine which stars are crossing the meridian at local midnight, by using a sundial to find true North and true South, and an hour glass to find the moment halfway between sunset and sunrise. If you are able to determine the moment of local midnight to within a couple of minutes, your results will give you the date.

Govert Schilling
Utrecht, The Netherlands

When calculating times longer than a few million years, you can use the rotation of our Galaxy. The Sun revolves around our Galaxy once in about 200 million years, and this lets you measure time by locating the position of the centre of our Galaxy relative to distant galaxies. Unfortunately, this is hard to do with optical telescopes. To locate the centre exactly you will need a radio telescope. Radio telescopes are bulky, so you will have some difficulty fitting one into your time machine.

For times longer than about 100 million years, you can use the recession of galaxies by measuring their angular diameters. The angular diameter of the nearby galaxy M33 was about

10 per cent larger 100 million years ago. For times longer than, say, 500 million years, you can use distant galaxies that recede as the Universe expands; their angular diameters were about 10 per cent larger a billion years ago.

Martin Clutton-Brock
University of Manitoba, Canada

Deep Breath

Q ***Is it true that every time we take a breath of air or swallow a mouthful of water, we consume some of the atoms breathed or swallowed by Leonardo da Vinci (as I read in a children's science book in 1960)?***

Steve Moline
Wentworth Falls, New South Wales

A We do indeed breathe in a considerable number of molecules that once passed through Leonardo's lungs and, unfortunately, Adolf Hitler's or anyone else's for that matter. The calculation is not too difficult and is as follows.

The total mass of the Earth's atmosphere is about 5×10^{21} grams. If we take air to be a mixture of about four molecules of nitrogen to one of oxygen, the mass of 1 mole of air will be about 28.8 grams. One mole of any substance contains about 6×10^{23} molecules. So there are about 1.04×10^{44} molecules in the Earth's atmosphere.

A single mole of any gas at body temperature and atmospheric pressure has a volume of about 25.4 litres. The volume of air breathed in or out in the average human breath is about 1 litre. So we can assume that Leonardo da Vinci, in one breath, breathed out about 2.4×10^{22} molecules.

The average human takes, say, 25 breaths per minute, so during his 45-year lifetime (1452 to 1519) Leonardo would have breathed out about 2.1×10^{31} molecules. So, about 1 molecule in every 5×10^{12} molecules in the atmosphere was breathed out by Leonardo da Vinci.

However, because we breathe in about 2.4×10^{22} molecules with each breath, there is a pretty good chance we breathe in about 4.3×10^{9} molecules that Leonardo breathed out. In fact, you can also show in a similar way that you probably breathe in about 5 molecules that he breathed out in his dying gasp.

Of course, there are some pretty crude assumptions involved here in order to arrive at the conclusion. We assume that there has been a good mixing of Leonardo's molecules with the rest of the atmosphere (quite likely in 500 years), that he didn't recycle some of his own molecules, and there is no loss from the atmosphere due to later users, combustion, nitrogen, fixation, and so on. But there is still scope for a considerable loss of molecules without it affecting the main point of the calculation.

By knowing that the number of molecules in the hydrosphere is 5.7×10^{46} molecules, similar calculations can be made for water. These show that a mouthful of liquid contains about 18×10^{6} molecules that passed through Leonardo during his lifetime. So, in addition to breathing in his breath, there is also a pretty good chance of picking up some of Leonardo's urine in every glass of water that you drink.

Peter Borrows
Epping, Essex

The law of conservation of matter ensures that atoms are constantly being recycled in the Universe. Gravity ensures that most of those on the Earth stay there. Some of the atoms floating around were breathed by da Vinci, although the number of these atoms compared to all those in the Earth's atmosphere would make them pretty few and far between.

However, considering the length of time in which, say, the dinosaurs inhabited the Earth, you can be pretty sure that every breath you take contains what was once part of one or more of these creatures, and that every apple you eat has many atoms that were part of an animal, even a human. Of course this could have some very worrying implications for vegetarians.

Glenn Alexander
Wollongong, Australia

A This question provides some food for thought for homeopaths. There is a very high probability that a cup of water contains a few homeopathic molecules that are effective in countering every illness we may have, and at no cost.

Lassi Hyvarinen
Le Vesinet, France

F-FACTOR

Q ***The following made its way into the* New Scientist *office...***

Please read this sentence and count the Fs:

FINISHED FILES ARE THE RE-
SULT OF YEARS OF SCIENTIF-
IC STUDY COMBINED WITH THE
EXPERIENCE OF YEARS.

How many did you see? On first reading most

people see only three. However, the answer is actually six. Why is this?

Rick Eraho
Cleckheaton, West Yorkshire

The fact that most people can only see three Fs instead of six would only be strange if reading was entirely phonetic. In reality, several methods are used to get meaning from print, and the most common of these has little to do with individual letters and their sounds.

A reader becomes familiar with the shape of many words, particularly short, common words like 'of'. These shapes are memorized and the reader no longer sees the words as made up of separate sounds. So, when you read the sentence, you spot the Fs in the longer, less familiar words but not in the three occurrences of 'of'.

Sam Hill
Exeter, Devon

An intelligent seven-year-old or a proofreader would read six Fs, because they have learned to give all words equal values.

As we learn to read faster, we select the most important words, permitting our brains to fill in the gaps. The faster we wish to read, the more words we must skip. It is quite possible to read simple narrative at faster than 600 words per minute, with comprehension. Fact-packed scientific text obviously has to be read more slowly.

A fast reader must concentrate on the most important words, usually nouns and verbs. Adjectives and adverbs come next, with modals, articles, pronouns and prepositions, and so on coming last. Experienced readers take less notice of the least important words which in English (as in most other languages I know) tend to be short

words. So all the little words like 'of' are skipped, and their Fs are omitted from the count.

I can only speculate about reading speeds in Chinese or Japanese where writing systems are quite different.

Valerie Moyses
Banbury, Oxfordshire

I tried this on a work colleague, without using the same layout of words and characters. When I rewrote it two occurrences of 'of' were at the end of lines, and my friend spotted these Fs but missed the remaining one. So I would assume that the way they are placed (in the middle of the line) has an effect on their noticeability.

Obviously, someone who does not understand English would have no problem spotting the Fs. It would be interesting to see how someone who has English as a second language responds to this.

Andrew MacCormack
Bristol

I found out by reading an optical illusions book *Can You Believe Your Eyes?* by J. Richard Block and Harold E. Yuker that most people think there are three Fs because they do not notice the Fs in 'of'. This is because the F in 'of' is pronounced as a 'V', so the brain doesn't recognize it as an F.

Bryn Hart (aged 10)
Kelmscott, Australia

Great shame was felt, not only at seeing only three Fs but at still seeing three after reading the answer. Then my wife sat down and announced that she saw six on first reading.

I am an English teacher and my wife a maths teacher and there lies the difference. The reader tries to understand the passage and casually ignores the word ‘of’ which is repeated three times. The mathematical mind does exactly what the question asks—it counts the Fs. My wife sees a number of letters but I see a phrase and therefore ignore the actual task.

The splitting of words with hyphens is also important because it puts pressure on the reader to understand the text and to follow the words, diverting attention from the real task.

Tom Sweetman
High Peak, Derbyshire

Your earlier correspondent will be pleased to know that English is my second language (Taiwanese is my first) and I am due to take maths A level exams this summer. My F count on first reading was three, and it was still three after seeing the answer. I only realized that the additional Fs were in ‘of’ when I spelt it all out aloud. In fact, I only realized as I recited the second ‘O. F.’.

It is my belief that your background has no bearing on whether you spot the Fs or not. My training as a systems analyst should help me to concentrate on the task in hand (counting the Fs). As to why I missed them, two explanations remain. Either it was too late at night or I am obviously not suited to doing a maths degree and should contemplate a change of course.

Alex Lu
Edinburgh

As a postscript I submit the following:

THE
SILLIEST
MISTAKE IN
IN THE WORLD

When trying to persuade a class of seven-year-olds of the need to reread what they had written, I put the above legend on the blackboard. My best readers immediately read it as 'The silliest mistake in the world' and I replied, 'You've just made it.' It was only when we reached the slower readers that the extra 'IN' was discovered. At this point the headmaster came into the room, glanced at the board, and read the message aloud, but incorrectly. A chorus from the class greeted him: 'You've just made it, sir.' The F and V confusion that was suggested by one of your correspondents cannot be the explanation here.

Douglas Boote
Stockport, Cheshire

STRANGE NATURE

MIRROR IMAGE

Q ***Why is an image in a mirror inverted left to right but not top to bottom?***
Kishor Bhagwati
Lausanne, Switzerland

A The mirror does not reverse images from left to right, it reverses them from front to back relative to the front of the mirror. Stand facing a mirror. Point to one side. You and your mirror image are pointing in the same direction. Point to the front. Your mirror image is pointing in the opposite direction to you. Point upwards. You both point in the same direction. Now stand sideways on to the mirror and repeat. You are now pointing in opposite directions when you point sideways. Place the mirror on the floor and stand on it. This time you point in opposite directions when you point upwards and your upside down image points downwards. In all cases the direction reverses only when you point towards or away from the mirror.
Hilary Johnson
Malvern, Worcestershire

A The answer stems from the fact that a reflection is not the same as a rotation. Our bodies have a strong left–right symmetry, and we try to interpret the reflection as a rotation about a central vertical axis. We imagine the world in front of the mirror has been rotated through 180 degrees about the mirror's vertical axis, and it has arrived behind the mirror where we see the image. Such a rotation would put the head and feet where we expect them, but leave the left and right sides of the body on opposite sides to where they appear in the reflection.

But, if instead we imagine the world to have been rotated about a horizontal axis running across the mirror, this would leave you standing on your head, but would keep the left and right sides of your body in the expected positions. The image would then appear top–bottom inverted, but not left–right.

So whether you see the image as left–right inverted or top–bottom inverted—or for that matter inverted about any other axis—depends upon which axis you unconsciously (and erroneously) imagine the world has been rotated about. If you lie on the floor in front of a mirror you can observe both effects at once. The room appears left–right reflected about its vertical axis, while you interpret your body as being left–right reflected about a horizontal axis running from head to foot.

Peter Russell
London

Actually, a mirror does not invert at all. Look at your face in a mirror: the left side appears on the left and the right on the right.

Now look at someone else's face without a mirror. It has been inverted because of the rotation necessary to turn and look at you: their right side is on your left. They could equally well turn to look at you by standing on their head, in which case you see their left on your left, but now the top of their head appears at the bottom. We don't normally do this because it's not very comfortable.

Try this experiment. Write a word on a piece of paper and hold it up to a mirror. You automatically rotate it about a vertical axis and it appears in the mirror inverted left to right. It is this rotation which inverts the image, not the mirror.

Try the experiment again, and this time when you hold the paper up to the mirror, rotate it about a horizontal axis. The word will be inverted top to bottom.

Alan Harding
Stansted, Essex

A The problem is caused by the way we visualize the mirror image. We imagine ourselves standing on a carousel, which has done a half-turn to put us where we see the image—that is, in the mirror. We see that the top and bottom of our bodies in the mirror image are in the same place, but left and right are reversed.

If instead of a carousel we used a ferris wheel to rotate ourselves, and imagined ourselves strapped upright in the seat, we would see a different result. When the wheel does a half-turn, the mirror image now has left and right in the correct places, but top and bottom are reversed.

The trouble is that we are incorrectly using rotation for these experiments, when, in reality, the mirror reflects front-to-back. Because this is a difficult thing to do with our body, we mentally substitute rotation, which doesn't quite fit what we see. Generally, we prefer to keep top and bottom correct, so we see a left to right reversal in a mirror, although we could see top to bottom reversal if we wished.

David Singer
San Francisco, California

Wet, wet, dry

Q ***What is the difference between the concrete used in dry construction and the one used***

underwater that solidifies in totally wet conditions?
Roberto Velez
Canberra, Australia

A There is no difference between the concrete used above and underwater. The setting of concrete is a hydration process, and can only occur in the presence of water.

If 'dry' concrete is allowed to dry out before the curing process has finished, the concrete will be weak. For this reason, concrete slabs with a large surface area are often covered with a polythene sheet to slow down evaporation. Concrete underwater only requires protection to prevent it being washed away or diluted. In certain circumstances special concretes may be required, such as sulphate-resistant concrete for use in seawater.
Andrew Parker
Tanga, Tanzania

MIDDAY MADNESS

Q ***There are more hours of daylight after noon than before it, particularly in summer. Does this mean midday is in the wrong place?***
Dean Sherwin
Reading, Berkshire

A Midday refers to the moment when the Sun crosses the local meridian, which is one of the imaginary lines joining the North and South Poles at 90° to the equator. If you set your watch so that noon occurs as the Sun crosses the meridian, the day will be of equal length either side of noon.

However, this system would mean that you would need to reset your watch if you travelled

only a short distance east or west. To avoid the confusion this would cause we use time zones —areas throughout which we say that the time is the same, irrespective of the actual meridian. Time zones are nominally 15° wide but in practice they vary in size and shape because of political, geographical, and practical considerations. The difference between the local meridian and the time setting of your time zone can be quite apparent if you live near the edge of an oddly shaped time zone.

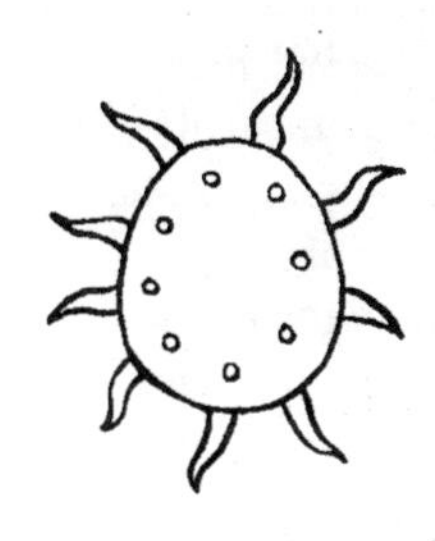

David Eddy
Perth, Australia

The development of time zones is usually attributed to the development of the railway system in the US, which runs predominantly east to west. Until the advent of the railways, most towns followed local time and the clock midday was well aligned with the solar midday. Then trains began to travel from town to town so quickly that the constant compensation for different local times caused timetable difficulties, prompting the development of time zones.
Keith Anderson
Kingston, Australia

The standard time used in Britain is based upon the Greenwich meridian and the latitude of your correspondent in Reading is almost the same as that of Greenwich, but his longitude is 1° West. Sunrise, local noon, and sunset therefore occur about four minutes later than at Greenwich, and Reading's local time is actually four minutes later than the standard clock time used in Britain.

This means that, at Reading, the duration of daylight after noon, as shown on the clock, is on average longer than the duration of daylight before noon. East of the Greenwich meridian, afternoon daylight is, on average, shorter than morning daylight. At Greenwich the difference between the duration of morning and afternoon daylight, averaged over a year, is zero.

On any particular day, the difference between the duration of morning and afternoon daylight depends, not only upon the latitude and longitude of a place, but also upon the equation of time. This is the difference, in time, between the mean Sun, which gives us clock time, and the true Sun. It is caused by the eccentricity of the Earth's orbit around the Sun, and the tilt of the Earth's axis in relation to its orbital plane. The equation of time varies during the year from minus 14 minutes to plus 16 minutes, and it is the main reason for the difference between the time that you calculate by looking at a sundial and that shown on a clock. There is also a slight difference between morning and afternoon caused by that day's portion of the Sun's annual movement around the ecliptic.

The combination of the above effects can create a difference between morning and afternoon daylight of more than half an hour at Reading.

None of this means, however, that midday is in the wrong place, merely that the standard time system, whose simplicity and uniformity are essential for communication, is necessarily an approximation to the Sun's complex apparent motion.

The further lengthening of the afternoon daylight, and shortening of before-noon daylight, during the months of British Summer Time, are of course, the intended result of the forward movement of clocks by one hour.

David Le Conte
The Astronomical Society of Guernsey

Midday Greenwich Mean Time is only the middle of the day at the Greenwich meridian. If you are west of Greenwich, as Reading is, the Sun rises later and sets later, so 1200 GMT will be earlier than the midpoint between sunrise and sunset. The Sun travels 360° in 24 hours, or 15° in one hour. Hence, as I write this in North London (0° 10′ West), 1200 GMT is 24 seconds before midday, but if I lived in Swansea (3° 56′ West), 1200 GMT would be nearly 16 minutes before midday.

Using Central European Winter Time, 1200 is 6 minutes before midday in Berlin (13° 30′ East), but it is more than 50 minutes before midday in Paris (2° 15′ East).

The most extreme example is Lisbon in Portugal (9° West), which has recently adopted Central European Time—during summer, 1200 is two and a half hours before midday.

Nigel Wheatley
London

SEALED IN LIGHT

Q ***When I am opening some types of self-sealing envelopes I notice that there is a purple fluorescent effect within the gum. It only lasts for a very short time, but can be repeated if I reseal the envelope and pull it apart again. What causes this effect?***

Stewart Duguid
Edinburgh

A The coloured glow is a form of chemiluminescence. Separating the gummed surfaces requires energy that breaks the attractive forces between the molecules of gum.

Presumably, the act of pulling apart the surfaces supplies excess energy to the gum molecules that lifts them into an excited state. As they decay back to their normal state the energy is released in the form of visible light. The difference in energy between the excited and ground states defines the wavelength and hence the colour of the light produced; in this case purple.

This phenomenon is different from fluorescence, where light (often ultraviolet) is absorbed and then re-emitted at a longer wavelength (in the visible spectrum). Fluorescence gives rise to 'Day-Glo' colours and the blue glow you might observe while drinking tonic water near one of the ultraviolet lamps often found in nightclubs.

Paul Wright
Peel, Isle of Man

A similar effect can be seen when stripping off a length of electrical insulating tape. I first noticed this about thirty years ago and the discovery came, by coincidence, shortly after an explosion in a coal mine. The last people to go down the mine before the explosion had been a crew of electricians.

I wondered if the electricians had been using insulating tape and so I wrote to the authorities questioning the possible danger of insulating tape as a source of ignition. However, I received a reply which stated that the effect was well known, but that there was insufficient energy in the sparks to ignite methane in the mine.

Mike Gay
Canada

As your previous correspondents mention, I noticed this glow—on Royal Society of Chemistry envelopes in my case—and wondered about ignition of flammable atmospheres. As I facetiously pointed out to the Society, members of the Royal Society of Chemistry often open envelopes in environments of lower ignition energy than that of methane.

More recently, there has been an explosion attributed to essentially this cause, or at least to peeling off an adhesive label. The entry in future editions of *Bretherick's Handbook of Reactive Chemical Hazards* may run something like this:

Adhesive labels
Tolson, P. *et al.*, *J. Electrost* 1993; 30: 149
A heavy-duty lead-acid battery exploded when an operator peeled an adhesive label from it. Investigation showed that this could generate ›8 kV potential. Discharge through the hydrogen/oxygen headspace consequent upon recharging batteries caused the explosion. The editor has remarked vivid discharges when opening Royal Society of Chemistry self-adhesive envelopes.

P. Urben
Kenilworth, Warwickshire

Beaming

Q **Why does the same side of the Moon always face the Earth? It can't be because by coincidence its rotation on its axis is exactly that of one orbit around the Earth. There must be a better explanation.**
Roger French
New Hampshire

A Despite appearances, the Moon is not perfectly spherical, and its denser core (and its centre of gravity) is not truly central. The lighter crust is thinner on the nearside, and thicker on the farside. This is somewhat analogous to the position of the yolk inside an egg.

Tidal forces exerted by the Earth would have rapidly halted any initial rotation, leading to the current, stable orientation with the lightest hemisphere pointing away from the Earth. Such spin-orbit coupling is universal among satellites in the Solar System, the only confirmed exception being Hyperion, a moon of Saturn, which is tumbling in its orbit, perhaps as a result of a fairly recent collision.
Storm Dunlop
Chichester, West Sussex

This feature of locked or synchronous rotation is not completely understood, but is caused by the effects of tides raised in one orbiting body acting on the other. These slow the rotation of the Moon until its rotational period matches the orbital period.

The Moon raises tides in the Earth, both in its oceans and (to a lesser extent) in its solid land surface, causing bulges in the Earth itself. These tidal bulges do not lie directly below the Moon, but lag some way 'behind' it as the Earth rotates. The gravity in the raised mass of the bulge below the Moon pulls on it and the lag causes this pull to be at an angle. This affects both the rotation and the orbit of the Moon.

The Earth, likewise, raises tides in the Moon, causing the Moon's surface to flex. This flexing causes friction, generating heat and rotational drag, again contributing to the slowing of the Moon's rotation.

The Moon is also not uniform in density or shape. There are several areas of increased density called mascons, and a significant bulge on what is now the far side. It may be that the position of the bulge and/or the mascons are responsible for the orientation we now have

—somewhat like a weight on a wheel controlling its final rest position.
Andy Hain
No address supplied

ROCK AROUND THE EARTH

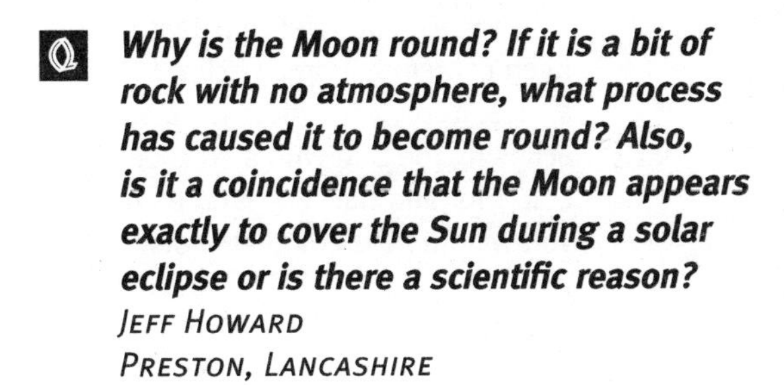

Q ***Why is the Moon round? If it is a bit of rock with no atmosphere, what process has caused it to become round? Also, is it a coincidence that the Moon appears exactly to cover the Sun during a solar eclipse or is there a scientific reason?***
Jeff Howard
Preston, Lancashire

A The moon is round because it is large. This has nothing to do with the erosional effects of atmosphere. Gravity tries to shape any object into a sphere, because every point on a sphere's surface is as close to the centre of gravity as it can get.

For smaller objects, like bricks or asteroids, the internal strength of the rock is stronger than the gravitational force acting on the surface, so their irregular shape persists. On the Earth, for instance, it is impossible for mountains to be higher than about 15 000 metres (only a small fraction of the Earth's diameter). Yet, on Mars, a planet smaller than the Earth, the highest mountain is 25 000 metres high, because the gravity of Mars is weaker. What's more, most large planetary objects were formed hot and partially molten, easing gravity's sphere-shaping job.

There is no scientific reason for the fact that the Sun and the Moon are apparently equal in size.

It is sheer coincidence. The surface of the Earth is the only planetary surface in the Solar system from which a perfect natural solar eclipse can be observed, with the Moon exactly covering the surface of the Sun.
Govert Schilling
Utrecht, The Netherlands

A Watch the Sun and Moon long enough (over millions of years) and you will see that the coincidence you refer to is temporary. If you had been contemplating an eclipse from the back of a triceratops, you would have seen the Moon eclipsing the Sun with a considerable overlap. Loiter for another few hundred million years, and you will find the complete solar eclipse a thing of the past. Tidal effects are moving the Moon further from the Earth into slower orbits and lengthening our day in the process. This will not stop until the Moon is in geostationary orbit and the Earth's day length equals the month length.
Jon Richfield
Dennesig, South Africa

UNDER THE RAINBOW

Q ***I have twice driven through the end of a rainbow where it meets the ground, although I am told this is impossible. I was surrounded by the rainbow's colours and there were many other drivers to witness that it can happen. Can anyone explain this?***
Peter Herbert
Loughborough, Leicestershire

A It is not possible to drive through a rainbow, but I would be interested in more details of

what happened, as there may be another meteorological phenomenon that could explain the experience.
Malcolm Brooks
The Met Office Press Office, London

The easy way to reach the end of a rainbow is to wish for a pot of gold next time you place a superfluous tooth under the pillow. If you use this method, do not forget to carry a spade at all times.

In order to be seen to be at the end of a rainbow, get a friend with an intercom to direct you to where the rainbow appears to touch down. In order to see yourself at the rainbow's end, arm yourself with binoculars and make sure that your friend holds a large mirror at the correct angle.
William Pearce
Warrington, Cheshire

Malcolm Brooks of the Met Office states that this experience is impossible, but is it really so rare to find the end of a rainbow? I walk to work past the fountains of the Serpentine in Kensington Gardens in London, and on two separate occasions have observed the end of a rainbow. There was, however, demonstrably no pot of gold there.
Barry Rigal
London

On 7 April 1994, following a storm, I was driving east from Swindon to Reading. A rainbow had formed and appeared to end in the middle of the road ahead. I placed a camera on the dashboard and made a number of exposures, up to a point where I appeared to drive through

the end of the rainbow. If this is impossible, what is happening?
Dennis Mackie
Maidenhead, Berkshire

A It is impossible to reach the end of a rainbow, as the observer must be between the Sun and the rainbow. If the observer is moving through a shower of rain, the rainbow will keep pace and remain at the same distance from the observer. I have seen this innumerable times at sea. On some occasions the end of the rainbow appears to be just under the bow of the boat but, like an asymptote, you just never quite reach it. It would appear likely that Dennis Mackie drove through a thin band of rain and thought he had driven through the end of the rainbow. However, it would have disappeared just before he reached it, hence the myth of the pot of gold.
R. A. Neill
Alloway, Ayrshire

A I regularly produce artificial rainbows in the lab and lecture theatre. The theory is that because a rainbow is formed not at a fixed place but at a certain angle between your eye, the Sun, and the raindrops, it is impossible to reach its end.

In reality, as many of your readers have pointed out, it is possible to get very close to the end, particularly when the water drops are well confined, such as in a fountain or the spray from a hosepipe. What is happening is that the rainbow is being formed according to all the usual rules, but by water drops which may only be a few inches from the eye. In fact, if the water drops are close enough, it is possible to see two rainbows, one for each eye. In this case, you feel as if you are almost inside the rainbow and so indeed at its end.

The real test, though, is to carry on moving and then look behind you to see the rainbow that you have just passed through. You will find that it has disappeared, because it was never actually formed in a fixed position. Alternatively, get a friend to go round to the other side of the fountain. They will see nothing.

Colin Cartwright
University of Abertay, Dundee

I have had numerous rainbow experiences similar to those described by your correspondents, and in other circumstances, such as near waterfalls. The rainbows that one can apparently approach almost to their ends are associated with local spray. A transition from the general rainbow with which we are all familiar can sometimes occur very smoothly when driving.

Clearly, a good rainbow appears only where there is rain or fine spray floating in the air ahead, with strong sunlight shining continuously onto it. Also, for the best transition effect, it is important that the direction of travel is such that the rainbow's end—seen at about the level of the horizon—is more or less directly ahead. Then, anyone approaching a vehicle from behind, and overtaking it, will drive through its spray, gaining an illusion of approaching and entering the end of a rainbow.

Andrew Smith
Woodford Green, Essex

A rainbow often appears closer on wet roads because spray from vehicles provides a sufficient number of water droplets to give a rainbow below the horizon. Because spray droplets are smaller than raindrops, the colours here are less intense.

J. East
Menai Bridge, Gwynedd

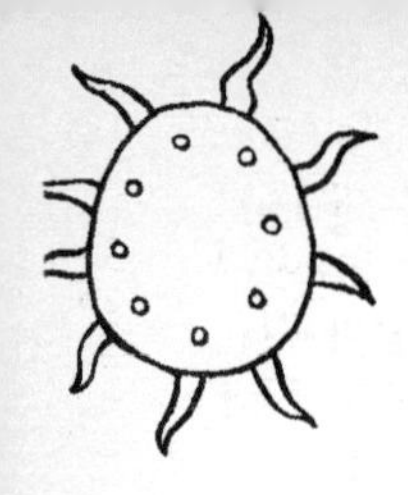

CLEAR DAY BLUES

Q ***Why (on a clear day) is the sky blue?***
Jaspar Graham-Jones
Southampton, Hampshire

A The sky is blue because of a process called Rayleigh scattering. Light arriving from the Sun hits the molecules in the air and is scattered in all directions. The amount of scattering depends dramatically on the frequency, that is, the colour of the light. Blue light, which has a high frequency, is scattered ten times more than red light, which has a lower frequency. So the 'background' scattered light we see in the sky is blue.

This same process also explains the beautiful red colours at sunset. When the Sun is low on the horizon, its light has to pass through a large amount of atmosphere on its way to us. During the trip, blue light is scattered away, but red light, which is less susceptible to scattering, can continue on its direct path to our eyes.
Rick Eraho
Cleckheaton, West Yorkshire

A The sky is blue because of a process known as Rayleigh scattering. According to classical physics, an accelerated charge emits electromagnetic radiation. Conversely, electromagnetic radiation may interact with charged particles causing them to oscillate. An oscillating charge is continually being accelerated and hence will re-emit radiation. We say that it becomes a secondary source of radiation. This effect is known as the scattering of the incident radiation.

The atmosphere is, of course, composed of various gases that together form air. We may treat each air molecule as an electron oscillator. The electron charge distribution of each molecule presents a scattering cross-section to the incident radiation. This is essentially an area upon which the incident radiation must fall for scattering to occur. The amount of scattered radiation will depend upon the magnitude of this cross-section. In Rayleigh scattering the cross-section is proportional to the fourth power of the frequency of the incident radiation. Sunlight is composed of various visible frequencies ranging from low frequency (red) to higher frequency (blue) light. Because it is of a higher frequency than other visible components, the blue part of the Sun's spectrum will be scattered more strongly. It is this scattered light that we see and so the sky appears to be blue.

Incidentally, we are also able to explain why sunsets are red. When the Sun is close to the horizon its light must travel through more atmosphere. The blue light will be scattered strongly whereas red light, because it is of lower frequency, is less prone to scattering and so is able to travel straight to the observer.

D. Roberts
Physics Department, University of Sheffield

JELLY ROLL

Q **Why does a jar containing Swarfega resonate so when bumped?**

Bruce Buswell
Bath

A Swarfega, like many materials, has both a viscous and an elastic side to its nature. It is a gel made up of a network of weak elastic bonds. These are easily broken under the shearing action caused by your fingers when using it to clean your hands. If these bonds are not broken down but are subjected to a force within their elastic limit (such as being bumped in a jar) they will store the energy and oscillate like a spring.

The period of oscillation is related to the bond energy and length. So when large networks of strong and relatively short-range bonds are struck, as in a metal anvil, you get high-pitched ringing tones. Weaker, longer bond networks like those in Swarfega give low-frequency natural harmonic oscillations when struck. This low resonance is quickly damped by the viscous component of Swarfega which, instead of storing the energy of a strike, dissipates it in forms such as heat and entropy.

Wayne Collins
Toddington, Bedfordshire

Swarfega is either a gel or a very viscous liquid (and there may be a phase change within the normal range of room temperatures). It is somewhat unusual in that most common substances of that nature have high internal friction losses, and you normally get a thud by striking the tin. The low internal losses of Swarfega suggest that the substance might, on the molecular scale, have some long-range structural order.

Because it is a detergent, its molecules will have an ionic end, which is attracted to water, and a fatty end, which is repelled by water. These molecules could form roughly spherical associations, with the fatty ends outwards and water in the centre. These would then slide easily over each other when the substance was grossly deformed and could resonate mechanically with low losses if the driving disturbance is of small amplitude. I recall that if some water is added to the Swarfega, the resonating effect is much reduced.

J. M. Woodgate
Rayleigh, Essex

SUN UP OR DOWN?

Q **_I have just been studying landscape scenes in a calendar and began to wonder if you can tell the difference between a sunset and a sunrise photograph. I once heard that the dust in the atmosphere causes a difference in colour but does anyone know if this is true?_**

Jim Royston
Bundoora, New South Wales

A At sunset the air is still warm. There is more dust, water vapour, pollution, and insect life in the air than at sunrise. However, although these can certainly change the colour of the visible sunlight, the main difference between the light at dusk and the light at dawn is the overall distribution of it, as artists and photographers will appreciate.

At sunrise in a graveyard the scene shows contrast: the shadows are bluer, slightly darker, and appear sharper than at sunset (I am ignoring the occasional effects of mist on the ground). The sky towards the east will still be quite cyan. At sunset, however, the scene is warmer and softer because the light from the west is coming from a broader area of the sky, diffused by the dust from the atmosphere.

I should emphasize that this is how it is in southern England, most of the time. Latitude, altitude, environment, and, of course, the weather have major parts to play.
Mike Akbari
Hove, Sussex

Sunrise and sunset are not created in identical ways and it is possible, in principle, to tell the difference between them by optical means alone. However, in practice this can be difficult.

The difference between sunrise and sunset is caused by differences in the compositions of night air and daytime air. An important example of such a difference involves the airborne concentration of aerosols—dust particles and chemical aggregates. Very fine aerosols, of a size comparable to the wavelength of light (in the range 1/10 000 of a millimetre) are the most efficient at producing deep red sunsets while large dust particles are more efficient at creating brilliant sky glow.

Diurnal variations in both types of aerosol are familiar to city and country dwellers alike. In cities, daytime aerosols come in the form of smog and dust from human activities while in the country, vegetation produces fragrant, blue-haze aerosols while wind can raise larger dust particles. In either case, the higher concentration of both types of light scatterers in the evening air can create deeper red sunsets and brighter sky glow. However, such effects depend on the height and breadth of the aerosol cloud and can be quite small for localized aerosol distributions.

Furthermore, natural variations in the aerosol content of the atmosphere caused by the weather can overshadow the visible effects of diurnal aerosols. So although there is a difference between sunrise and sunset, it is usually difficult to tell with the human eye alone.

Riccardo Brun del Re
Kingston, Canada

LIGHT STAR

Q ***If the sun is a star just like all the others, why does it appear yellow rather than white?***

John Berry
Newark, Nottinghamshire

A As a beam of light passes through the air, blue light tends to be scattered while the red and yellow continue to pass through. This is why the Sun looks red at sunset as the light passes through a great thickness of atmosphere, and also why the sky is blue.

If you put the beam back together, adding the light from the blue sky and the yellowish Sun, it would appear white. This addition of light covering the full visible spectrum is what happens when you look at a field of snow.

The same thing happens to starlight. But the cells in your eye need a large input of energy to detect colour, so dim stars seem colourless. If the sun were a distant star it too would look white.

Spencer Weart
Center for History of Physics, Maryland

A Contrary to popular belief, stars come in many different colours. This is because their colour is intrinsically linked to their temperature.

Young, hot stars will be white, while cooler, older stars will be red. Our Sun is a medium-sized star, burning at a medium rate. This is reflected in its yellow colour.

Sion Amlyn
Trefor, Gwynedd

A There are two main reasons why stars appear white, even though they really cover a wide range of colours from deep red (cool stars, less than 3 000 °C) to bluish-violet (hot stars, greater than 30 000 °C).

First, the human eye is poor at detecting colours at low light levels. For example, all cats look grey in moonlight. Hence, most stars, especially the fainter ones, appear white.

Secondly, many of the stars visible to the naked eye are genuinely white or bluish-white, being among the hottest and most luminous ones in our region of the Galaxy, such as Rigel in Orion.

Notable exceptions are Betelgeuse in Orion and Antares in Scorpius, which are highly luminous, cool red giant stars near the end of their lives, although these do appear distinctly red when compared to other stars.

Other stars of the same temperature (about 5 500 °C) and size as the Sun, or even cooler and smaller ones, are either too dim for their colour to be distinguished by the naked eye or cannot be seen at all without a telescope.

Paul Hatherly
University of Reading

WARM FRONTS

Q ***I have often been told that 'it will get warmer when it snows'. This seems counter-intuitive, yet the weather in Edinburgh was freezing until a couple of days ago when it started to***

snow. Is this coincidence or does snowing actually warm up the air?

Edward Myse
Edinburgh

A The reason it seems warmer after it has started to snow is because it is warmer. Snow can fall when ambient temperatures are between 1 and 3 °c. So the atmosphere needs to warm up before the snow can start to fall.

Charlie Ferrero
No address supplied

A Snowing does not cause the warmth, the relative warmth causes the snow. During very cold weather the capacity of the air to contain moisture is greatly reduced. What water vapour remains is deposited as frost. At the same time, evaporation is reduced and therefore air humidity remains low. In these conditions snow cannot form.

Warmer air can hold more moisture before saturation point is reached and the water vapour begins to condense out. So when warmer air moves into cooler areas not only do we feel the increase in temperature but the excess vapour precipitates as snow.

Kevin Lowe
Wolverhampton, West Midlands

THE PHYSICAL WORLD

DRIP DRY

Q ***If I alight from the bus and it is raining, I tend to run for my door, in the belief that I will arrive home less wet than if I walk. However, I have heard that the same number of raindrops will strike me whether I run or walk. Is this really the case?***

MARK HAINES
LONDON

A The volume of space swept by the body between bus and door is identical. Therefore, assuming a constant rate of deluge, the number of falling raindrops swept (below the crown of the head) is the same. However, the number of raindrops falling directly on to the top of your head is proportional to the time spent exposed to the rain, so running reduces this component.

But running will deposit all the swept component in a shorter time. This will produce greater apparent wetting since normal evaporative drying has less time to work. So, for light showers, with small swept and falling components, walking is probably preferable. We make this complex decision completely unconsciously, while also taking into account the likelihood of the rain becoming harder or lighter, the distance we have to travel, and our ability to run.

It would be interesting to confirm this theory by filming pedestrians, recording the rate of rainfall, and relating the latter to the point at which the former begin to run.

MIKE STEVENSON
MILLOM, CUMBRIA

A If the walker is in the Lake District, where horizontal rain is common (and always opposes the direction of travel), then it is recommended that they move as quickly as possible because the volume of drops swept out through the rain is now determined by the relative velocity of rain and walker multiplied by the time taken on the journey. Indeed, if the rain is moving over the ground at speed vr (opposing travel) then the walker, moving at speed vp, will be $1 + vr/vp$ wetter than in vertical rain by the time they reach shelter. By running to keep up with the rain (defined as $vp = -vr$) it is theoretically possible to stay dry.

Martin Whittle
Sheffield, South Yorkshire

A The following (from memory) is attributed to one D. Brown of York:

When caught in the rain without mac
Walk as fast as the wind at your back
But when the wind's in your face
The optimal pace
Is as fast as your legs can make track.

Matthew Wright
University of Southampton

Chinese Puzzle

Q ***The Great Wall of China is commonly cited as being the only non-natural object visible from space. For an object to be visible when viewed from above, the eye must be able to resolve it in two dimensions.***

The Great Wall of China is immensely long, but very narrow. If the eye is able to resolve its width from space, many other objects such as the Great Pyramid of Cheops should be large enough to be seen in two

dimensions, despite having a much smaller total area.

Is the ability of the eye to resolve objects in the smaller of the two dimensions affected by the magnitude of the larger (and if so why) or is the claim made for the Great Wall of China incorrect?

A. R. MacDiarmaid-Gordon
Sale, Cheshire

A The claim is incorrect. It is well known as one of the most widely believed urban legends, perhaps second only to the famous one about mass suicide by lemmings. A person with perfect eyesight is able to resolve up to about one minute of arc without binoculars or a telescope. The Great Wall of China is, very approximately, 6 metres wide. This means that it is not directly visible above an altitude of about 20 kilometres, or just over twice the height of Mount Everest. Even if its shadow is taken into account, this would only make it visible, in places, up to perhaps about 60 kilometres at the most. Because of atmospheric drag, this is still below the height necessary for a stable spacecraft orbit.

There are, however, many man-made objects which are visible from outer space, the largest being the Dutch polders or reclaimed land. Cities too can be seen at night because of the bright streetlights.

D. Fisk
Ipswich, Suffolk

A It is well known that the human eye can pick out long objects much more easily than short ones, so the Great Wall of China is certainly a candidate for being visible from the Moon. However, the wall is, in places, a broken-down edifice and is often scarcely visible on the ground, never mind from

space. H. J. P. Arnold, photographic expert and skilled astronomer, has studied this problem and concludes that seeing the wall from the Moon is a physical impossibility.

Neil Armstrong of Apollo 11 has stated that the wall is definitely not visible from the Moon. Fellow astronaut Jim Lovell (Apollos 8 and 13) made very careful observations and says that the claim is absurd. Jim Irwin (Apollo 15) has said that seeing the wall is out of the question.

Photographs from uncrewed probes do show that the route of the wall is sometimes shown by sand that is blown on to the windward side, but that the wall itself is not visible. The end, perhaps, of yet another legend.

Robert Brown
Ashby de la Zouch, Leicestershire

Forked Frolics

Q ***Why does lightning fork and what is the diameter of a bolt of lightning?***
Michael Lee
London

A Lightning usually brings the negative charge from a thunderstorm down to the ground. A negatively charged leader precedes the visible lightning, moving downwards below the clouds and through air containing pockets of positive charge. These are caused by point discharge ions released from the ground by the thunderstorm's high electric field.

The leader branches in its attempt to find the path of least resistance. When one of these branches gets close to the ground, the negative charges attract positive ions from pointed objects, such as grass and trees, to form a conducting path between cloud and ground. The negative charges then drain to ground starting from the bottom of the leader channel. This is the visible 'return stroke' whose luminosity travels upwards as the charges move down. Those branches of the leader that were not successful in reaching the ground become brighter when their charges drain into the main channel.

Photographs of lightning often overestimate the channel width because the film can be over-exposed. Damaged objects that have been struck by lightning show channel diameters of between 2 and 100 millimetres.

R. Saunders
Atmospheric Physics Group,
Manchester University

Rising cold

Q ***Heat is supposed to rise. Why then, is it colder on top of a mountain than in the valley below?***
J. M. Allen
Burley, Hampshire

A Heat doesn't rise. Low density air rises if surrounded by air of higher density. Air that is heated by contact with warm ground becomes less dense and therefore rises. As it rises, it mixes with cooler air above and cools to a point where it stops rising. Air at the top of a mountain makes little contact with the ground and is therefore cold. Air in the valley below makes a great deal of contact and is therefore warm.
Mike Martin
Newtown, Australia

A First, there is a much greater air volume at altitude, because the planet is a sphere: like an onion skin, each layer of atmosphere covers a larger and larger area. The ground heat therefore dissipates into this vast heat sink. Secondly, there is no lid on the atmosphere (ignoring for our purposes the greenhouse effect).

Heat therefore escapes from this sink into the endless void by way of radiation.

Mike Eadie
Petersfield, Hampshire

The lower region of the Earth's atmosphere, the troposphere, is warmed by radiation from the Earth's surface. Therefore the warmest part of the troposphere is that nearest to the ground.

The stratosphere above absorbs incoming radiation from the Sun. The temperature here therefore increases with increasing height. So if you climb 40 kilometres above the Earth's surface you would find that it gets much warmer again.

Rosemary Gray
Beverley, East Yorkshire

The previous letters all raise important issues but the fundamental reason is based on the gas laws. If air at a low altitude rises, it moves into an area of lower pressure. If there is no heat transfer its temperature falls. The air expands as it rises and this expansion does work against the surrounding air, taking energy from the rising air and cooling it.

So the dynamic equilibrium condition in the atmosphere is one of falling temperature with increasing altitude.

N. D. Whitehouse
Stockport, Cheshire

Air is a poor absorber of sunlight. The Sun heats the Earth which is a better absorber, and the Earth then heats the air close to it. Hot air rises because it is less dense than surrounding cooler air at the same pressure. As it rises, a mass of hot air expands because the ambient pressure is less. This expansion cools the air, so the temperature of a thermal bubble decreases with increasing altitude until it reaches equilibrium.

Alan Calverd
Bishop's Stortford, Hertfordshire

This phenomenon is explained by what is termed adiabatic expansion. Think of a balloon inflated at sea level which is then carried up a high mountain. As the atmospheric pressure drops, the balloon will expand because a given mass of gas will occupy a larger volume if it is at a lower pressure.

The work that expands the balloon against its elasticity and the pressure of the gas outside comes from the air inside it. As this air is allowed to expand it does work. This energy in turn is derived from the thermal energy (motion) of the air molecules inside the balloon and because of the work that it is doing the air cools.

This phenomenon will cool the air by approximately 9 °C per kilometre of altitude. Therefore, air at 20 °C at sea level would be in neutral equilibrium with air at 11 °C at 1 kilometre altitude. This is close to the temperature gradient that is seen in the lower 8 kilometres of our atmosphere, above which other effects dominate.

Dominic Gallagher
Oxford

The air pressure at any point is a measure of how much air lies above that point, so air pressure decreases with height. As air pressure decreases, temperature decreases; so temperatures will nearly always be colder on the top of a mountain than in the valley below.

The exception is when inversions occur, but this does not alter the basic premiss. If the air is heated at the Earth's surface, it will rise because it is less dense than the air around it—the warmer air is, the less dense it is.

As the air rises, its pressure decreases, and thus it cools, so again the air on the mountain will be cooler than in the valley. This is a simplification of the many processes that are happening in the atmosphere, but even so it is the most important.

Malcolm Brooks
The Met Office, Bracknell, Berkshire

A To describe the atmosphere as being like the skin of an onion, and therefore having more volume in its outer as compared to its inner layers, is conceptually wrong.

Because the Earth is nearly 13 000 kilometres in diameter and the top of the troposphere is only 13 kilometres above the surface, a more realistic analogy is that the parts of the atmosphere are more like the skin of an orange rather than the layers of an onion. Thus, the volume of air at a given height does not increase significantly with altitude and so this effect is insignificant in determining temperature drops.

Julian White
Chepstow, Gwent

IN A SPIN

Q ***What makes the Earth rotate?***

R. J. Isaacs
Barnet, Hertfordshire

A The Earth rotates simply because it has not yet stopped moving. The Solar System, and indeed the Galaxy, were formed by the condensation of a rotating mass of gas. Conservation of angular momentum meant that any bodies formed from the gas would themselves be rotating. As frictional and other forces in space are very small, rotating bodies, including the Earth, slow only very gradually. The Moon, a much smaller and lighter body, has effectively already stopped rotating because of the gravitational drag exerted by the Earth, and now always keeps the same face turned towards us.

GLYN WILLIAMS
DERBY

A Although the previous explanation is correct, the assertion that 'the Moon . . . has effectively already stopped rotating' could possibly be misleading. The Moon does rotate. The reason why it presents the same face to us is that its period of rotation is the same as its period of revolution around the Earth. This equality is the result of tidal friction. If the Moon did not rotate, any line through it, parallel to the orbital plane, would keep the same direction in space and the Moon would show us its far side during a complete revolution, as one can easily convince oneself by making a drawing on paper.

D. S. PARASNIS
DEPARTMENT OF GEOPHYSICS,
LULEÅ UNIVERSITY OF TECHNOLOGY, SWEDEN

STRAIT NOT NARROW

Q ***In the Ordnance Survey National Atlas of Great Britain (1986) it states that the English Channel widens at the Strait of Dover by 70 centimetres each year. Even if the figure is much smaller, surely any widening in some way affects the new tunnel. How is the widening allowed for?***

C. S. MOORE
HALIFAX, WEST YORKSHIRE

A In reference to the widening of the English Channel at the Strait of Dover, Ordnance Survey

confirms that the figure of 70 centimetres each year represents coastal erosion, not the movement of Great Britain relative to France.

There are very minor changes to the distance between England and France, which amount to approximately 1 millimetre every ten years. This is caused by the Southeast of England sinking at a slow rate.

When Eurotunnel and TML (the contractors) contemplated the design of the tunnel project, this fact was taken into account and 30 000 joints in the tunnel can comfortably take up the movement.

Colin Kirkland
Eurotunnel, London

MAD TIDINGS

Q ***Can anyone explain in simple and common-sense terms why there is simultaneously a high tide on both sides of the Earth?***

Pat Sheil
Sydney, Australia

A In considering the origin of tides we must disregard the Earth's daily rotation around its axis and concentrate only on the revolution of the Earth–Moon system.

This revolution takes place around the system's common centre of gravity, which is about halfway from the surface to the centre of the Earth, and causes every point in the Earth's interior or on its surface to describe a circle of radius equal to the distance of the common centre of gravity from the Earth's centre.

Therefore, at every point there is a centrifugal force of the same magnitude and in the same direction: away from the Moon, parallel to the line joining the Earth–Moon centres. This centrifugal force is distinct from the one caused by the Earth's rotation, which we are disregarding.

Every point of the Earth also experiences a gravitational force as it is pulled towards the Moon, the direction of this force is different for different points of the Earth.

The resultant of these two forces creates the tide-generating force. If we now consider two points on the Earth's surface, one directly below the Moon and the other on the far side, it turns out that the Moon's gravitational force at the near point is greater than the centrifugal force which, as we have seen, is away from the Moon.

The far point is farther away from the Moon by one Earth diameter and the Moon's gravitational force there happens to be smaller than the centrifugal force, so the net force on water at the far point is away from the Moon.

In most popular accounts, the simultaneous occurrence of tides at the two opposite points is explained by asserting that while the Moon pulls the water at the near point some distance, it pulls the Earth's body a little less.

But this explanation does not clarify why a system like that will not simply collapse under the mutual gravitational attraction between the Earth and the Moon.

D. S. Parasnis
Department of Geophysics, Luleå University of Technology, Sweden

A Ignoring the effects of other bodies, the centre of mass of the Earth and the centre of mass of the Moon are both in free fall, following orbits around the common centre of mass of the Earth–Moon system, which gravity and centrifugal acceleration precisely cancel out.

Over most of the Earth's surface, though, this cancellation is not precise, because you're either nearer to, or farther away from, the Moon, but still forced to orbit at the same rate as the Earth's centre of mass.

For the ocean on the side of the Earth facing the Moon, lunar gravity dominates centrifugal force, so water bulges towards the Moon.

On the opposite side, centrifugal force dominates, so water bulges away from the Moon. Both bulges produce high tides.

In effect, sea level—which would otherwise be spherical—is stretched along the Earth–Moon axis into an ellipsoid, and as any point on the Earth rotates into and out of either bulge, the local tides flow, then ebb.

Greg Egan
Perth, Australia

The simultaneous high tides on opposite sides of the Earth are a result of an imbalance between gravitational forces and centrifugal forces. Tides are caused by the gravitational interaction of Earth and Moon, and to a lesser extent the Earth–Sun interaction.

Although we think of the Moon as orbiting the Earth, in fact the Moon and the Earth both orbit their common centre of mass, which is close to, but not exactly at, the centre of the Earth. The centrifugal forces generated by the orbital motions of each body just balance the gravitational pull of the other body.

However, the balance is only exact at the centre of each body. On the side of the Earth nearest the Moon, the Moon's gravitational pull is slightly greater and the centrifugal force slightly less than at the Earth's centre, so water here is pulled out towards the Moon. On the opposite side of the

Earth, the gravitational pull is slightly less and the centrifugal force slightly greater, so here water is thrown outwards away from the Moon.

Mark Bertinat
Chester

Early Days

Q ***The shortest day of the year in the northern hemisphere occurs on 21 or 22 December, yet the earliest sunset is on 13 or 14 December and the latest sunshine occurs a similar number of days after the shortest day. Why is this?***

John Walker and Alan Whittle
Manchester

A Two properties of the Earth's orbital motion around the Sun give rise to the curious disparity between the dates of earliest sunset, shortest day (winter solstice), and latest sunrise. These are the eccentricity of the Earth's orbit, and the tilt of its equator to its orbital plane.

The combined effect of these is to vary the length of the day throughout the year. In some months, the interval between noons on successive days is slightly greater than 24 hours, while in other months, it is slightly less. The differences cancel one another out over the course of a year.

In December, near the northern winter solstice, the interval between successive noons is about 30 seconds less than 24 hours. Because this difference is greater than the daily change in sunrise and sunset times, it becomes the dominant effect, causing the observed separation of the dates of earliest sunset and latest sunrise.

A similar effect is seen in June, but the interval between successive noons is then only about 13 seconds less than 24 hours, so the dates of earliest sunrise and latest sunset are closer to the summer solstice—the longest day.

Fred Watson
Coonabarabran, Australia

A The main effect of this shifting of latest sunrise to several days after the winter solstice (the shortest day of the year) is an oscillation of the time of day when the Sun reaches its highest elevation. The time oscillates about noon in a sine wave with an amplitude of 8.8 minutes at latitude 45 degrees and with a period of six months. At the solstices and equinoxes, highest elevation is at noon.

Sunrises and sunsets each day occur equally before and after the time of the Sun's highest elevation. If we call those intervals morning and afternoon, then at the winter solstice the length of morning is at its minimum and so changes very little for a few days. But although the time of the Sun's highest elevation is noon at the solstice, that time gets later quite quickly and shifts the sunrise so that it falls later for a few days.

After that few days, morning begins to get longer and overwhelms the 8.8 minute sine wave so we get earlier sunrises again.

Why does the time of the Sun's highest elevation oscillate around noon? Mainly because the Earth's axis is tilted with respect to its orbit around the Sun. Fix a frame of reference at the centre of the Earth, aligned with the axis of the Earth, but rotating uniformly around that axis only once per year, not once per day. In that frame, the Sun generally goes up and down by 23 degrees. But it also goes side to side a little, and this gives

the oscillation in time of the Sun's highest elevation each day around noon.
Terry Watts
New Jersey

BOOM...HISS...

Q ***Presumably the big bang made a sound. How loud was it in decibels, say, and can we still hear it, if we strain our ears?***
Nicholas Buck
British Columbia, Canada

A Our ears can only detect sounds in the atmosphere (or other gas surrounding our heads). Space is an almost perfect vacuum so sounds detectable by our ears will not propagate in it. Radio telescopes can hear the big bang when they detect the background 2.7 K radiation of the Universe.
George Sassoon
Warminster, Wiltshire

A The big bang did not produce sounds in the conventional sense and the name is unfortunately misleading with its suggestion of a violent explosion (and accompanying noise). No one has yet come up with a memorable, descriptive alternative, although a term such as 'primordial expansion' might give a better sense of how the whole of our Universe (space, time, and matter) originated within an infinitesimal region at that instant. We also know nothing about the effects that the creation of our Universe had upon pre-existing or surrounding universes (if these exist), in which the laws of physics and physical constants may differ from our own.

In addition, sound waves are pressure waves which, by their very nature, propagate through a material medium and thus through space. But our own space only came into existence with the big bang, so in effect the sound of the big bang had nowhere to go.

There is a sense, however, as recently suggested by Alexander Szalay of Johns Hopkins University, in which we may study sounds produced just after the big bang. At this stage, the primeval Universe was dense enough for acoustic waves to propagate within it. Some 300 000 years after the big bang, the temperature dropped sufficiently for electrons and protons to combine to form atoms. The Universe became transparent, allowing radiation to travel freely and giving rise to what we now observe as the cosmic background radiation at about 2.7 K. At the same time, the acoustic waves were no longer able to propagate through space, and may have been preserved as density fluctuations that eventually gave rise to the large-scale structure of the Universe. This may account for the pattern of clusters of galaxies and voids that we observe today. If this theory is correct, studies of the minute fluctuations in the cosmic background radiation may reveal primordial structures of the same scale as those shown by modern galactic surveys.

Storm Dunlop
Chichester, West Sussex

BURNT OUT

Q ***Can a candle burn in zero gravity? Without convection the sphere of outgoing gaseous***

products will surely prevent any oxygen getting in.
Malcolm Thomas
Cranleigh, Surrey

A candle can burn in zero gravity. The flame is a diffusion flame which tends to a spherical shape in the absence of airflow. The burning rate is reduced drastically because combustion only occurs in a thin spherical shell where the outward diffusing fuel vapours meet the inward diffusing oxygen. The flame loses the yellow colour of incandescent carbon particles and becomes an almost invisible clear blue. To test this, candles were burnt on board the space station Skylab.
David Bryant
European Space Agency, The Netherlands

INSIDE MACHINES

HOLDING THE LINE

Q ***How do welded railway lines, with no gaps, cope with expansion and contraction in different temperatures?***
G. L. Cantrell
Exeter, Devon

A Older rails buckled in hot weather once they had expanded to fill the gaps between them. Welded rails are laid down under tension, by being stretched, so a rise in temperature causes only a drop in the tension. With no need for gaps, passengers have a smoother ride. But if the temperature continues to rise after the tension has dropped to zero the rail will buckle, just like the old type.
Barry Lord
Rochdale, Greater Manchester

A In the US, welded railroad tracks accommodate linear expansion by vertical movement. Watching a train pass on a hot summer, say, here on the prairie, you can see the track being pushed down against the ballast by the locomotive, and rise again after the caboose (or guard's van) has passed.

When the wooden sleepers are well secured to the rails, a railroad track functions as a beam. Because the stiffness of a beam is proportional to the square of its dimension in the direction of the force, the horizontal stiffness of the track is approximately 60 times the vertical stiffness, so practically all the expansion in length is accommodated by vertical waviness. In winter, the waves disappear and the track becomes flat.
Ross Firestone
Winnetka, Illinois

A Welded rails do have gaps but, unlike the older rails, they are not butt-jointed. They look rather like half of a closed point, where the two lengths of rail to be joined are both machined at an acute angle, and the pointed end of one rail laid against the start of the machining on the other rail. The two rails can slide one against the other with expansion or contraction. Because the two are held together no gap appears, and the way they are laid ensures that the inside edges remain the same distance apart.

Richard Downs
Higher Denham, Buckinghamshire

I like the spoof answer to this question, but the two serious answers printed were incomplete.

Continuously welded rails were adopted because the traditional joints had to be dismantled twice a year for inspection and lubrication. A rail is quite capable of withstanding the compressive forces caused by heating, but the danger is the stress will relieve itself by buckling. Concrete sleepers are too heavy to permit the track to buckle in a vertical plane but expansion does take place in that direction, as your correspondent suggests. However, if a track gang lifts the track when it is in compression it will immediately buckle. Additionally, track everywhere at all temperatures sinks slightly into the ballast as trains pass.

In Britain, track has to cope with a temperature range of −7 °C to 49 °C. Given a wide shoulder of ballast outside the sleeper ends, the track will only withstand buckling over about half this range. Continuously welded rail is thus fastened down at temperatures of between 13 and 21 °C.

In the early days of railways, track laid in colder weather was de-stressed by waiting until the ambient conditions gave the right temperature, loosening the fastenings, and running a locomotive over it slowly to ease it into position before refastening it firmly again to the sleepers. Subsequently the technique was changed, and flame guns were used to heat the rails to the correct temperature when they were laid. That, in turn, has now given way to mechanical stretching.

On tight curves, even a good bed of ballast cannot stop buckling, and traditional fishplated joints are still used.

P. Semmens
York

KEY PROBLEM

Q **_Why are the rows on a calculator or number keypad arranged with the lowest numbers at the bottom, when we normally read from the top downwards? And why are telephone keypads arranged the other way, with the lowest numbers at the top?_**

A Mechanical adding machines, based on rotating wheels, always have the 0 button adjacent to the 1 button. By convention, most old adding machines had the numbers increasing in value from the bottom and this may be a hangover from when the machines had levers on the wheels rather than buttons. When the numbers were put on to a pad arranged as a three by three grid with one left over, the order of the numbers, as far as possible, kept the same.

On a rotary telephone dial the zero is adjacent to the 9 because a zero in the telephone number is signalled by 10 pulses on the line. When telephones acquired push buttons in a grid the ordering of the buttons was carried over from the old telephone dial.

Nicko van Someren
Cambridge

HEAVY METAL

Q ***During a recent thunderstorm, I left the room in which my music centre resides. On my return a few minutes later it had apparently turned itself on and was playing a CD. The chance of human (or indeed animal) intervention was very small, and my wife asserts there was a lightning flash while I was out of the room. Could the lightning have turned it on?***

Colin Lindsay
Chichester, West Sussex

Many readers wrote to say they had similar experiences during thunderstorms, so we can be fairly sure the phenomenon is real. The exact mechanism is less certain because a lightning strike causes many forms of electromagnetic radiation and can affect a sensitive CD player in more than one way—Ed.

A The probability of lightning causing this is quite high in electronic equipment such as hi-fi. For any system that can be switched on remotely with an infrared or radio frequency remote control unit, the power supply must provide power to the infrared or radio receiver at all times—even when it is apparently switched off by the equipment's power switch.

When lightning strikes, considerable electromagnetic energy is generated over a wide frequency range, which means that lightning can often be heard in AM broadcast radio receivers. This electromagnetic energy can induce small currents in the equipment, sufficient to trip some of the logic gates, which the infrared or radio receiver interprets as a signal from the remote control to switch on. New electromagnetic compatibility rules were introduced recently to make equipment less susceptible to this.

With equipment where it would be dangerous for this to happen, such as electric power tools, the power is switched off by a mechanical mains switch, not the electronic hi-fi one. Such equipment is then far less likely to switch on accidentally unless it suffers a direct strike with sufficient energy to weld parts of the wiring or switch together.

David Kirkby
Department of Medical Physics,
University College London

My CD player has also switched on during a storm. It is not a pleasant experience to be alone in a house with lightning flashing and thunder rolling outside, and the lights flickering and then the CD player unexpectedly turning itself on.

This phenomenon is not caused directly by the lightning, but by the power supply being cut off for a very short space of time, then coming back on. The same effect can be achieved by switching the power off at the socket and then on again very quickly.

Catherine Stevenson
St Bees, Cumbria

Lightning often induces pulses of electricity in power cables. It can start electronic devices, stop them, or make them do the weirdest things. When a device requires a coded signal, the pulse from lightning will not usually fool it. However, most of these receivers are used only for starting other components, using a simple pulse. If the lightning's pulse can bypass the receiver, it may directly switch on that other component without ever giving the code.

Of course, lightning also may induce a much heavier pulse which can fry the whole system, so it is a good idea to turn off your computer or any other valuable sensitive devices and unplug them when there is a storm in the offing.
Jon Richfield
Dennesig, South Africa

LIVE WIRE

Q ***Where does the force come from when you are thrown horizontally across a room after touching a live electrical connection? I thought there was a reaction to every action, but there is no obvious push from the electricity.***
John Davies
Ahmadi, Kuwait

A The force comes from your own muscles. When a large electrical current runs through your body, your muscles are stimulated to contract powerfully—often much harder than they can be made to contract voluntarily.

Normally the body sets limits on the proportion of muscle fibres that can voluntarily contract at once. Extreme stress can cause the body to raise these limits, allowing greater exertion at the cost of possible injury. This is the basis of the 'hysterical strength' effect that notoriously allows mothers to lift cars if their child is trapped underneath, or allows psychotics the strength to overcome several nursing attendants.

When muscles are stimulated by an electric current, these built-in limits don't apply, so the contractions can be violent. The electric current typically flows into one arm, through the abdomen, and out of one or both legs, which can cause most of the muscles in the body to contract at once. The results are unpredictable, but given the strength of the leg and back muscles can often send the victim flying across the room with no voluntary action on their part. Combined with the unexpected shock of an electrocution this feels as if you are flung, rather than flinging yourself.

The distance people can involuntarily fling themselves can be astonishing. In one case a woman in a wet car park was hit by lightning. When she recovered she found herself some 12 metres from where she was struck. However, in this case there may also have been some physical force involved from a steam explosion as water on her and the area in which she was standing was flash-boiled by the lightning. She survived, though she was partially disabled by nerve damage and other injuries.

A common side effect of being thrown across the room by an electric shock, apart from bruising and other injuries, is muscle sprain caused by the extreme muscle contractions. This can also damage joint and connective tissue. Physiotherapists, chiropractors, and osteopaths might consider asking new patients if they have ever been electrocuted.

Being thrown across the room can save your life by breaking the electrical contact. In other cases, particularly where the source of the current is something they are holding, the victim's arms and hand muscles may lock onto it. They are unable to let go and, if nothing else intervenes, they may die through heart fibrillation or electrocution.

I recall what may be an apocryphal account of a poorly earthed metal microphone causing a rock singer to be involuntarily locked on to it. Unfortunately, writhing on the floor while screaming incoherently was not entirely unusual during his shows, and it was a while before one of his road crew figured out that something was amiss and killed the power.

Roger Dearnaley
Abingdon, Oxfordshire

It is interesting to consider why the subject is thrown across the room rather than freezing in a tetanic posture. It is because some muscle groups dominate others. Compare this with the muscle effects seen in a stroke victim, where, if the stroke is severe enough so that no cerebral control is present over one side of the body, the arm is held flexed (that is wrist bent with fingers pointing to the wrist, elbow bent so that the forearm meets the upper arm) and the leg extended (knee straight, ankle extended so that the toes point to the ground).

This is because without cerebral control, the spinal cord reflexes cause all muscle groups to be active, including both components of any bending and straightening muscle pairs. The dominance of one muscle group over another produces the effect described.

Therefore, if any electrical charge triggers all muscle groups the imbalance in 'bending and straightening' muscle pairs produces the force that is required to throw the person across the room.

It's not at all recommended, but I have heard that if you touch a conductor carrying a current using the back of your hand it is safer than the palm because the resultant muscle spasm does not force you to grip the conductor, producing a continuing electrocution.

There is always the effect on the heart to consider too, but that is another matter.

John Parry
Cowling, North Yorkshire

DOT TO DOT

Q ***Many of the envelopes that I receive through the post have been stamped with a row of blue dots by the Royal Mail. Presumably this assists the automatic sorting process. Does reusing these envelopes confuse the sorting machinery in any way?***

VASSILI PAPASTAVROU
NO ADDRESS GIVEN

A The blue dots printed on envelopes by the Royal Mail are there, as suggested, to assist in the automated sorting of the mail. These dots are in the process of being replaced by a series of red lines which fulfil the same purpose.

There are either one or two rows of marks. If there is a row printed at the bottom of the envelope, this is a machine-readable representation of the address. If there is a row in the middle or near the top of an envelope, this is a machine-readable number which uniquely identifies that letter within the Royal Mail sorting system.

The marks are actually a phosphorescent material which is applied to the letters using ink-jet printers. A phosphorescent ink is used in preference to other materials because it gives an excellent signal to noise ratio if it is illuminated with ultraviolet light and viewed some fraction of a second later.

The questioner asks whether reusing an envelope printed with these marks would confuse the sorting machinery in any way. The answer is no. All the Royal Mail's new automated coding and sorting machines include readers to look for and interpret the rows of codes. If a reused envelope

enters the letter-sorting system, the existing code marks will be recognized and the letter sent for manual sorting. The downside though is that it costs the Royal Mail more to sort this letter than if the envelope were new.

Paul Barton
Head of Technology, Royal Mail, Swindon, Wiltshire

PULLING POWER

Q **_To escape the Earth's gravitational field, a spacecraft has to travel at escape velocity. Why can't a less rapid ascent be employed? If a rocket-powered device has enough thrust to lift its weight, surely it will eventually make it to space?_**

Graham Drake
Leighton Buzzard, Bedfordshire

A The concept of escape velocity applies only to unpowered projectiles, not powered rockets. Unfortunately, the definitions of escape velocity given in many textbooks do not make this clear. Obviously, a rocket rising vertically at a low speed would eventually reach space if it kept going at the same speed. But, in practice, rockets run out of fuel and become unpowered projectiles. If they have not reached escape velocity when this happens they will not escape the Earth's gravity.

Peter Lafferty
Wadhurst, East Sussex

A A spacecraft needs energy, not velocity, to leave the Earth's gravitational field. You can leave the Earth as slowly as you like, providing you do enough work against gravity along the way.

Suppose a spacecraft has a weight W on the launch pad. If the thrust from the rockets is greater than this, the spacecraft will move upwards. As it does so, it gains gravitational energy. If that energy gain is at least WR (where R is the radius of the Earth), gravity can never pull that spacecraft back to the surface.

The least efficient way of launching a spacecraft is to have the thrust slightly greater than the weight. The spacecraft moves up very slowly and runs out of fuel before it gets very far.

The most efficient strategy is to launch the spacecraft very quickly, giving it kinetic energy. It can then be left to coast the rest of the way, slowing down as it gains gravitational energy. The spacecraft will have enough kinetic energy if its launch velocity is at least 11 kilometres per second.

However, the high acceleration involved in such a rapid launch, to say nothing of the heat generated by friction with the atmosphere, means that a compromise strategy is needed. A constant acceleration of 30 metres per second per second gets the spacecraft out of the Earth's atmosphere in about 30 seconds with a speed of about 1 kilometre per second. About five minutes after launch the spacecraft will have travelled 1500 kilometres and will reach escape velocity. Earth's gravity will have dropped only by about a third. Nevertheless, the rockets can now be turned off.

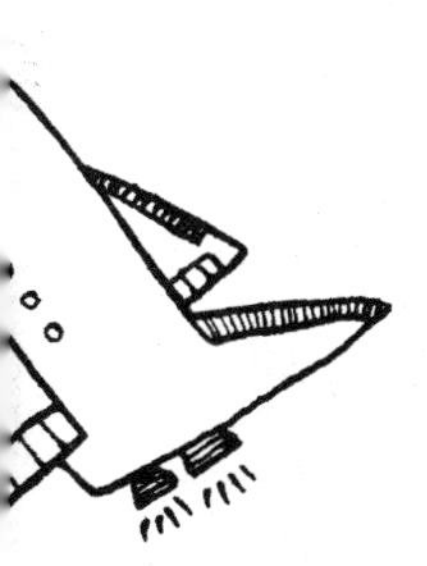

Michael Brimicombe
Aylesbury, Buckinghamshire

Q ***Why does NASA's space shuttle roll over on to its back within seconds of launching in Florida? I thought that being 'right side up'***

would help the astronauts in an emergency situation. I checked the NASA Web site for information on this but it waffled on and on and eventually lost the point altogether.
PAUL LEE
YORK

The shuttle waits to clear the launch tower before rolling. Then it rolls so that the angle of attack between the wind caused by passage through the atmosphere and the chord of the wings (the imaginary line between the leading edge and the trailing edge) is at a slightly negative angle.

This causes a small amount of downward force and alleviates structural loading. It is important that the wings are protected like this because they are one of the most delicate parts of the vehicle.

Following the roll, the new position allows the shuttle to carry more mass into orbit, to achieve a higher orbit with the same mass, or to change the orbit to a higher or lower inclination than would be the case if it didn't roll.

The rolled position also allows the crew to fly a less complicated flight path if they need to execute one of the more dangerous manœuvres, such as the 'return to launch site manœuvre', and it also improves the ability of ground-based radio antennae to communicate with the S-band radio antennae that are used on the shuttle.

Additionally, the crew can see the horizon, which is a helpful, though not mandatory, part of piloting any type of flying machine.

Finally, the shuttle is also oriented so that the body is at a lower angle to the ground, usually with the nose facing to the east. This allows the engine thrust to add velocity in the correct direction to achieve Earth orbit.

All this begs the obvious questions: why isn't the launching pad oriented to give this attitude to begin with and why does the shuttle need to roll to achieve that attitude? The answer is that the launching pads were left over from the days of the Apollo Moon launches and, to avoid having to construct new pads, the shuttle has to fit in with the old configuration by straddling two flame trenches: one for the solid rocket motor exhaust and the other for the space shuttle main engine exhaust.

George Curran
Hampton, Virginia

TAKING THE PLUNGE

On a recent flight from New York to Cleveland the plane dropped suddenly during a thunderstorm. Both the air steward and my reading glasses shot to the roof of the plane. The plane then stopped as quickly with a resounding bang and the steward fell on top of the passengers. How could we fall so much faster than purely under the influence of gravity and leave the steward behind, as it were? Does this happen often? How did we come to such an abrupt stop (seemingly harder than landing on a solid runway) and how far can the plane fall before it starts to break up?

A. J. Parr
Corwen, Clwyd

In straight and level flight, the lift force produced by a plane is exactly equal to its weight. Most of this lift is generated by the wings, which are set at

a slight angle to the oncoming airflow. If this angle increases, the aircraft produces more lift than its weight and it will climb. A decrease in this angle will produce less—or negative—lift, and the aircraft will descend.

When an aircraft flies into turbulent conditions, it may experience regions of rapidly rising or descending air. The motion of the air in front of the aircraft changes the angle at which the wings meet the oncoming wind and this changes the amount of lift produced. As a passenger, you experience this as a series of uncomfortable bumps.

If a particularly severe down-draught is encountered, the wings may produce negative, or downward, lift. The aircraft will then go into a curved downward flight path, which is the beginning of what a stunt pilot would call an outside loop. During this manœuvre, objects inside the aircraft will experience negative g and gravitate towards the cabin roof.

However, such strong downdraughts are often followed by equally strong updraughts and it is likely the aircraft will swiftly return to positive g flight. There is often a loud bang as the aircraft structure adjusts itself.

Large planes are designed to the standards of the Joint Airworthiness Requirements (JAR)—part 25. These state that an aircraft must be able to survive a vertical gust of 17 metres per second without exceeding the flight envelope limits. For a typical large transport aircraft these limits are +3.8 g and –1 g, where g is the acceleration due to gravity.

JAR also states that an aircraft must withstand an impact with the ground at a vertical velocity of 3 metres per second. This means that a well-designed plane would break into pieces if it was dropped to a hard surface from a height of just under half a metre. Fortunately, most pilots are able to touch down with vertical velocities considerably less than 0.5 metres per second.

Bill Crowther
University of Bath

A Because turbulence is a common problem which causes a few injuries to stewards and passengers, all modern aircraft have weather radar to warn the pilot of rain and thunderstorms ahead so they can be avoided. Modern aircraft and major airports have what is known as predictive wind shear equipment which enables the pilot to know exactly the whereabouts of severe downdraughts.

Terence Hollingworth
Blagnac, France

DRIVING RANGE

Q ***When running very low on petrol while driving on a motorway, what speed and gear would generally give the best mileage per gallon in order for you to reach the nearest service station?***

Julian Fletcher
Manchester

A On a motorway, speed affects fuel economy in two ways: through energy loss due to friction of components and through energy loss due to wind resistance. Generally, as a proportion of total energy loss, losses due to component friction are inversely proportional to vehicle speed, whereas losses due to wind resistance are directly proportional to speed. Although other factors such as the slope of the road come into

play, achieving maximum fuel economy is dependent upon finding a compromise speed at which these losses are reduced to a minimum.
James Walsh
Hemel Hempstead, Hertfordshire

The best figures will be achieved by engaging the highest gear ratio and driving at the lowest possible speed, in that gear, without labouring the engine. Between 65 and 110 kilometres per hour there is an approximately linear decrease in kilometres per litre as speed is increased. Hence, the slower you drive the less fuel is consumed per kilometre. The highest gear ratio will give the greatest number of road wheel revolutions for the minimum number of engine revolutions.
Neil Hunt
Maidenhead, Berkshire

Competitors on fuel economy drives use the highest possible gear and the lowest allowable speed. This minimizes drag forces and avoids wasting power that is used to simply turn over the engine and transmission. However, the rules of such competitions preclude coasting and there is a minimum average speed.

Without these restrictions, the most economical method is to pick a minimum and a maximum speed (say 65 and 95 kph) and accelerate at full throttle from the lower to the higher, then allow the car to coast until it has slowed to the lower speed, then accelerate again, repeating this process for the whole journey. In this way, the engine always operates at maximum efficiency.

The two speeds chosen depend to some extent on the drag coefficient (which can be substantially lowered by overinflating the tyres so they are very hard) and the range over which the engine develops its maximum torque. Always depress the throttle pedal gently, because a sudden flooring pumps extra fuel into the system, enriching the mixture unnecessarily. I have tried this method (on long night journeys when the roads were empty), and it can save up to 40 per cent on fuel consumption.

Graham Saxby
Wolverhampton, West Midlands

How far you will get in a car running out of fuel depends on the engine size, the speed, and the fuel. Fuel consumption (litres per kilometre) is optimal for the following combinations: 1600 cc engine running on diesel at 60 kph; 1600 cc on petrol at 70 kph; 1300 cc on petrol at 80 kph. Top gear is the best for all these combinations and the speeds are for engines in perfect condition. Extra wind resistance from items such as suitcases and bicycles on the roof will tend to reduce the optimal speed.

Jeremy Hodge
London

When driving at a constant speed, the lowest consumption is obtained at minimum speeds. The system used by drivers in mileage marathon competitions is called 'impulse–acceleration–impulse'. Accelerate up to 30 kph in fourth gear, then cut the ignition and change the gear to neutral. Let the speed go down to between 8 and 12 kph and start the engine again, repeating the operation. This method should not be used if your steering wheel is locked when you cut the ignition.

Jorge Moure
Buenos Aires, Argentina

Careful reading of the road ahead, coupled with measures such as accelerating briskly down hills and throttling back to travel up them can increase your range by as much as 10 per cent. As a further measure, and if you have the requisite amount of bravery/stupidity, you could sit a few feet behind a large vehicle to minimize air resistance. Just make sure it isn't mine.

K. J. Steward
Kettering, Northamptonshire

Don't try the techniques described in these last three letters on public roads—Ed.

DECEPTOR VANS

Q ***How does a TV detector van work? Surely a TV, being a receiver, emits very little electromagnetic radiation and so is indistinguishable from any other electrical appliance. Do TVs emit their own special signal? Or are the detector vans simply bogeymen used to scare us into paying our licence fee?***

Alexander Campbell
Edinburgh

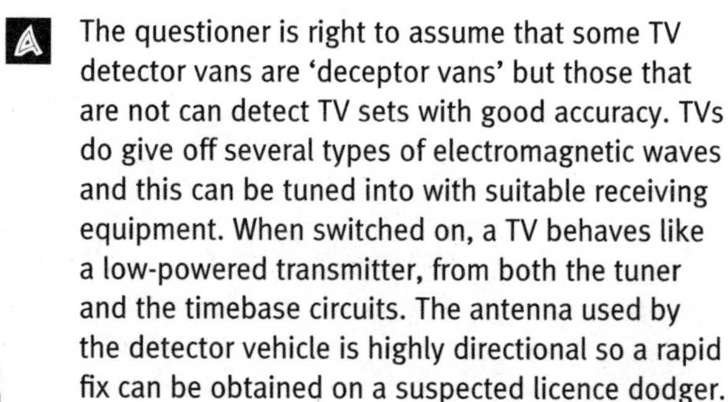

A The questioner is right to assume that some TV detector vans are 'deceptor vans' but those that are not can detect TV sets with good accuracy. TVs do give off several types of electromagnetic waves and this can be tuned into with suitable receiving equipment. When switched on, a TV behaves like a low-powered transmitter, from both the tuner and the timebase circuits. The antenna used by the detector vehicle is highly directional so a rapid fix can be obtained on a suspected licence dodger.

More usually, the authorities receive details of TV sales from the dealer after purchase and they

check them against their licence records. If you do not have a licence you may get a visit just in case you possess an unlicensed set.
Frank O'Brien
Harlow, Essex

In common with most radio receivers, televisions are superheterodyne receivers, in which the incoming high-frequency signal is mixed with a local oscillator signal to produce a lower, fixed-frequency signal—which is known as the intermediate frequency.

This is amplified, filtered, and detected so that the signal processing circuitry does not need to be retuned to receive different channels and can be designed to have a precise gain-frequency profile and very high rejection of out-of-band signals. In the case of a TV, accurate separation of the sound, vision, and colour signal components is also made possible.

Only the very first stage of the receiver has to accept very high-frequency signals over a wide band, and only the local oscillator needs to be tuned to receive different broadcast channels. It is tuned to the channel frequency plus the intermediate frequency, and the mixer generates the difference frequency, which is the intermediate frequency.

British colour televisions receive signals in the 470 to 860 megahertz UHF band, and their local oscillators operate between 510 and 900 MHz to generate an intermediate frequency of 39.5 MHz.

Although great care is taken to shield the local oscillator and mixer circuits, and to isolate the aerial from the mixer, some of this signal is radiated—mostly from the aerial—and can be detected by detector vans. Because the local oscillator frequency is always 39.5 MHz above the channel being received, the detector van can tell which channel you are watching.

A TV is also pretty noisy at other frequencies: some of the intermediate-frequency signal leaks out, and there is considerable radiation from the timebase scanning coils. These are driven by a pulsed signal at 14.625 kilohertz and so splatter characteristic higher-frequency harmonics into the ether (mostly as short-range magnetic signals); they can easily be detected with a long-wave radio near the TV.

Andrew Ward
Wotton-under-Edge, Gloucestershire

CUE CLUE

A few seconds before ITV or Channel 4 starts to broadcast adverts some televisions display a small black and white rectangle in the top right corner of the screen. Why do they do this and could it be used to stop video recorders from recording the adverts?

Stephen Phillips
Nottingham

The small squares appearing on the top right of TV screens on ITV and Channel 4 just before commercial breaks are called cue dots. They consist of vertical black and white stripes, to make them visible on any background. They are not locked to the picture, which is why they sometimes seem to spin around.

Their main purpose is as a visual warning to operators who are controlling video machines with manual rolls of tape. They appear 5 seconds before the time when the next machine should be switched on. This is usually to allow the operators to roll the adverts appearing in programme breaks. If cue dots appear during the programme, they have been used for editing purposes.

The ones used by the BBC appear in the top left corner, coming on and off twice to give a 5 and a 15 second roll, for older 2 inch tape machines. Film, tape, and cassette machines now stabilize within 3 seconds of being switched on, so 5 seconds is enough. Cue dots have been replaced at night by audio tones sent over telephone lines, which start automated machines remotely.

Cue dots don't appear on all TV receivers because some have been adjusted to make the raster (and hence the picture) larger than the screen.

Because the cue dot is part of the jumble of picture information, it would be hard to use it to avoid recording commercials on a VCR. The digital information sent among the picture frames may be more useful, but it would require complex circuitry. One problem would be that not all breaks are the same length—especially when programmes are live—so restarting your machine would be difficult. Another is that the tape and video heads would wear quickly if paused.

Anyway, you are supposed to watch commercials avidly, spend lots of money, and make someone else rich.

Peter Coen
Witton Gilbert, County Durham

If the signal is always the same it might be possible for video recorders to detect the pattern (or more importantly, when it ends). However, advertising breaks would have to be of a constant length, because there is no such signal when programmes resume. There is also a chance that, on rare occasions, the video recorder might be stimulated by the actual television picture itself.

Matthew Babb
Birmingham

HOUSEHOLD SCIENCE

SMELL FROM HELL

Q ***Why is it that, whatever they contain, dustbins always smell the same?***

Rodri Protheroe
Colchester, Essex

A The source of the smell is most probably caused by bacteria and fungi feeding on the organic matter in the rubbish. It will be most noticeable if the bin is in a warm and damp place.

The smell will not always be exactly the same, but it will be more characteristic of the different organisms than on the type of food they consume. The smell you get from penicillin mould growing on an orange is just the same as that from penicillin mould grown in a laboratory culture. It is pungent, characteristic, and very common. Analyses of household rubbish have detected very pathogenic bacteria, including *Pasteurella pestis*, the bacterium responsible for causing bubonic plague. So don't sniff too hard.

Cary O'Donnell
Welwyn, Hertfordshire

A I was pondering this question while taking out rubbish and I realized that dustbins do not smell the same. A bag containing foodstuffs will inevitably be ripped open by local cats unless protected by a dustbin, but a bag without food is not. It would seem that, although the bags smell similar to humans, they are noticeably different to cats.

As for why they smell similar, that would be because they invariably contain similar objects. However, garden refuse, for example, smells nothing like kitchen refuse, which in turn smells nothing like bathroom refuse.

Stewart Ravenhall
Newport Pagnell, Buckinghamshire

SLICE CRISIS

Q ***What is the irritant that causes eyes to water when slicing onions? Is there any way to prevent this?***
Stephen Mitchell
Redruth, Cornwall

A Onions and garlic both contain derivatives of sulphur-containing amino acids. When an onion is sliced, one of these compounds, S-1-propenylcysteine-sulphoxide, is decomposed by an enzyme to form the volatile propanthial-S oxide, which is the irritant or lacrimator.

Upon contact with water—in this case your eyes—the irritant hydrolyses to propanol, sulphuric acid, and hydrogen sulphide. Tearfully, the eyes try to dilute the acid. However, it is these same sulphur compounds that form the nice aroma when onions are being cooked.

To prevent watering eyes, I would suggest one of the following: stop using onions (but you would lose the tasty aroma); wear goggles (you would look slightly silly); slice the onion under water (you will wash some of the aroma out); before slicing the onion, wash it, and keep it wet.
Bernd Eggen
Exeter, Devon

A To attempt to reduce the severity of watering eyes you must allow the maximum possible time for the irritant to disperse before it comes into contact with the eyes. The most obvious way of achieving this is to stand as far away from the onion as your arms will allow. It also helps if you are not standing over the onion, but back from it.

Another way of reducing the amount of irritant reaching your eyes is to breathe through your

mouth. This means that instead of creating a current of air which flows up to your nose and onwards to the eyes, carrying the irritant with it, the air is either directed into the lungs when breathing in or forced away from the face when breathing out.

In order to ensure that you breathe through your mouth, hold a metal spoon lightly between your teeth. There will be space for the air to enter and escape and while our mouths are open we breathe preferentially through them, rather than our noses. I find that holding the spoon upside down works best, although I don't know what scientific reason there could be for this.

C. Burke
Farnham, Surrey

I have found that wearing contact lenses prevents eye irritation when chopping onions.

Elaine Duffin
Keighley, West Yorkshire

A slice of lemon should be placed under the top lip while slicing. One does not look particularly attractive but it does prevent the eyes from watering.

Sheila Russell
Staines, Middlesex

I suggest the old tip of holding a sugar cube between your teeth to absorb the irritant. It does work, as do sulphur matches, though very few people use these now.

Michel Thuriaux
Geneva, Switzerland

Hold a piece of bread—say a quarter of a slice —between the lips as you slice onions. This was taught to my family in Tanzania in the

early 1960s by our cook, Victor Mapunda, from Malawi.
John Nurwick
London

See also the question on p. 147—Ed.

QUESTION OF CLASS

Q ***At the risk of appearing a philistine, many people suggest letting red wine breathe before drinking it to improve the flavour. Wouldn't it be quicker to pour it into a cocktail shaker, shake for 10 seconds and let the bubbles subside?***
Chris Jack
London

A Wine is left to breathe to allow the volatile and aroma-bearing substances to start evaporating, so that we may enjoy the bouquet. Shaking a drink is completely different. An agitated drink incorporates gas, letting oxygen reach as much liquid as possible. This oxidized liquid provides a very different taste.

For some drinks this taste may be pleasant. However, if you oxidize wine you obtain vinegar, which, I suspect, is not the flavour you wish to taste. Therefore, there is a genuine reason for drinks being 'shaken, not stirred' or vice versa, depending on what you have in your glass.
Paul Mavros
Aristotle University, Thessalonika, Greece

A The reasons usually given for decanting red wines have changed during the past few years. This is because of two developments: one in

wine-making technology and the other in wine tastes.

The original reason for decanting was to separate the wine from organic particulates formed by precipitation, and aggregation from tartaric acid, tannin compounds, original microparticulates present in the pressed grape juice, and proteinaceous material that is formed during maturation of the wine.

Because these particulates are small to minuscule in size and of a density not much higher than the wine itself, Stoke's law predicts that they will sink back to the bottom only extremely slowly should they be suspended by careless motion of the bottle.

This is the reason for those magnificent mechanical decanting machines which allow precisely controlled tilting of the bottle to reduce suspension of particulates.

A very different reason for decanting lies in aerating the wine to hasten the release of the secondary elements of its nose. While traditional old wines may actually lose some of their olfactory elements through intense aeration and become stale quickly, decanting for aeration parallels the development of taste in younger wines or wines elevated in oak casks with associated different weighting of primary and secondary smells.

In Italy, where many progressive vintners have been experimenting with new assemblages and methods of elevation, decanting often means pouring the contents of a bottle straight down into a decanter, generating lots of chaotic turbulence with an intense mixing of air and wine.

In the hands of a self-confident wine waiter this process can look flamboyantly spectacular. As a logical development of this reason for decanting, some modern Italian glass decanters have a flattened shape that allows for the maximum air–wine interface giving further aeration.

Oliver Straub
Basle, Switzerland

A It is generally recognized that red wine should be drunk at ambient temperature and because it is often stored in a relatively cold room or location (near the floor) the most important aspect of the so-called breathing process is to raise the wine's temperature.

However, the ambient temperature in the UK is often a little low and red wine is usually best if drunk at about 30 °c. Placing a bottle of red wine in a microwave oven for 50–60 seconds (depending on the season) on high power will produce the required effect without having to resort to allowing the wine to breathe before consumption, but do not forget to remove the foil capsule and the cork. The alternative concept of shaking the wine in a cocktail shaker will result in the formation of various oxidation products including vinegar, which will have a negative effect on the flavour.

M. V. Wareing
Braintree, Essex

Only chemists drink red wine at a temperature of 30 °c. Our wine experts suggest a temperature of around 17 °c—Ed.

CREAM ON

Q ***One of the recommended ways of drinking the liqueur Tia Maria is to sip it through a thin layer of cream. If the cream is poured on to the surface of the drink, to a depth of about 2 millimetres, and left to stand for***

about two minutes, the surface begins to break up into a number of toroidal cells. These cells develop a rapid circulation pattern which continues even if some of the Tia Maria is sipped through the cream. How and why do these cells develop and what is the energy source?

GEOFFREY SHERLOCK
AMERSHAM, BUCKINGHAMSHIRE

This is a truly astonishing effect for which not a single reader has produced an explanation. 'Rapid circulation pattern' does not do justice to the series of eruptions that convulse the surface of the cream as the liqueur bursts through from beneath. Reach for a bottle and be amazed. Then sit back and work out why it happens. Extra toroidal cells can be generated by puncturing the surface of the cream with a skewer. Drinking off a little of the cream can also regenerate activity.

Additional data gathered in the *New Scientist* office may help. What is needed for the effect? The liquid must be dense enough to support a layer of cream. Substitute water or neat gin and the cream simply sinks to the bottom of the glass. The liquid must also contain alcohol. Substitute blackcurrant cordial for Tia Maria and the cream just floats there motionless and no cells form.

Provided the liquid is strongly alcoholic and dense, almost any mix will do. Gin and soy sauce is particularly effective—pour on the cream and the toroidal cells appear in seconds, even though the taste leaves much to be desired.

Could it be that the molecules of fat in the cream and those of the alcohol in the liqueur are immiscible and fighting a fierce battle at the surface of the glass? Do doughnut-shaped cells appear because toroids minimize the surface area

around a hole? What exactly is happening between the fat and the alcohol? And how can the circulation be sustained for so long? Any answers we print in a future edition of *New Scientist* will win their authors a bottle of Tia Maria—Ed.

TINTED GLASSES

Q ***What causes crystal glasses to cloud when washed in the dishwasher and can the process be reversed?***

Gail McKechnie
No address supplied

A Crystal glass will cloud in dishwashers because of the alkalinity of most dishwasher detergents. The high pH dissolves one or more of the constituents of the glass, leaving a microscopically pitted surface. As far as I know, there is no remedy.

Dan Taylor
No address supplied

A Crystal glasses owe their clarity to their very smooth surface. However, they can be softer than cheaper glasses. Because many dishwasher powders and liquids include abrasive agents, these can cause many minute scratches on your glasses and thus dull their once gleaming shine. Sadly, the effect is not reversible. To prevent this happening, either use a non-abrasive detergent in the dishwasher or, horror of horrors, wash your glasses by hand!

Steve Wooding
Sharnbrook, Bedfordshire

A Crystal glasses cloud when washed in a dishwasher because the strongly alkaline detergent etches and roughens the surface of the glasses. Glass is composed of a three-dimensional network of silicon dioxide, with various metal ions stuffed in the network's holes rather like a plum pudding.

Ordinary glasses contain sodium and calcium ions in these holes. Lead crystal has lead ions in addition and/or substitution. The detergent dissolves these ions to form hydroxides, which causes the etching. Glasses that have a high lead content are most susceptible to etching and should always be washed by hand, using soap or mild detergent, and towel dried. Cotton or linen towels have a mild polishing effect as bartenders often demonstrate.

The loss of metal ions can be reversed by an ion exchange in which the glasses are put in a molten salt bath containing the relevant ions.

However, it is not necessary to ion exchange the glasses to restore the lustre. This can be restored by either chemical or mechanical polishing. Chemical polishing uses buffered hydrofluoric acid, which requires elaborate safety precautions because it can cause wicked burns, blindness, chronic illness, and even death.

Mechanical polishing uses cerium oxide, which is considerably safer. Micrometre-size powder is mixed with water to form a slurry and rubbed over the surface of the glasses using a wool buff. A small hand grinder is convenient for powering the buff. The speed and pressure should be kept low to avoid heating and/or breaking the glasses.

Ross Firestone
Winnetka, Illinois

STICKY PROBLEM

Q ***Why does sticky tape when pulled from a roll quickly (at 10 millimetres a second) become almost transparent, but when pulled off***

slowly (at 1 millimetre a second) become opaque? Indeed, if while pulling tape off quickly, one pauses for a few seconds, a distinct line is left on the otherwise transparent tape. Can anyone explain?

David Holland
Broadstone, Dorset

The reason for the difference in behaviour lies in the response of the adhesive layer on the tape to the rate at which it is stressed. When the tape is peeled back slowly, the adhesive responds by forming long drawn-out strands between the two pieces of tape which break and fall back on to the tape, forming an opaque, rippled surface. These strands can be seen with the naked eye or with a hand lens.

When pulling off at higher speeds, instead of being able to stretch, the incipient adhesive strands formed break at much lower elongations and produce much less disturbance of the adhesive layer.

The difference arises because of the viscoelastic nature of the polymer which forms the sticky material. The material has a viscous component giving it some of the physical properties of treacle. It also has an elastic component which causes it to behave like a solid material such as metal in the form of a wire. When treacle is stretched it forms long strands almost never breaking, whereas metal wire has a comparatively low elongation and breaks when pulled. At low pulling rates, the adhesive is more like treacle and at high rates, more like the metal wire.

Ultimately, the behaviour is dependent on the time of relaxation processes at the molecular level. Because time is in some sense equivalent to temperature when considering molecular movements, it is interesting to see what happens when the tape is cooled in a freezer. Now, pulling at the lower speed produces a much more transparent region. Because there is not enough time for the long-chain molecules to unravel, the adhesive breaks in a brittle manner.

Stephen Hancock
Stockport, Cheshire

HITTING THE BOTTLE

Q ***If you remove the foil from a bottle of still wine, then hit the bottom of the bottle against something firm such as a tree trunk or the corner of a bar, the cork will slowly work its way out of the bottle. How does this happen?***

Nick Leaton
No address supplied

A Each time the bottle hits the tree a compression wave travels along the glass. This wave exerts a force on the glass and it decelerates.

So, shortly after the base has hit the tree, the whole bottle has stopped moving. This includes the cork. A similar compression wave travels through the wine, starting near the base of the bottle and moving towards the neck. As the wave travels down the neck of the bottle, its amplitude increases as the cross-section area of the wine decreases.

Eventually, the wave reaches the cork and tries to stop it moving towards the tree. However, because compression waves travel faster through solids (the glass) than through liquids (the wine), the cork has already stopped moving. So the energy in the compression wave is used to push the cork out of the bottle.

The method works best if the bottle is on its side, so there are no air bubbles between the cork

and bottle base. If you hold the bottle upside down you may end up with wine all over the floor.
Michael Brimicombe
Aylesbury, Buckinghamshire

A I once saw a spectacular demonstration of the force of a such a wave in water. In a martial arts demonstration the expert gripped the open top of a wine bottle full of water between his thumb and forefinger and then by slapping his other hand on to the top of the bottle he knocked the bottom of the bottle off.
John Goodier
London

A May I suggest that anyone who wants to try this wraps the bottle securely in several layers of towelling or similar, because on the one occasion I tried it, I finished up needing six stitches in my right forefinger and had to spend time picking up broken glass from my lawn.
Hugh Davies
Stagsden, Bedfordshire

One or two?

Q ***Expert advice says that you should use freshly drawn water every time you make a pot of tea or coffee. Why is this? What is wrong with water that has been boiled twice? Can anyone tell the difference?***
Ivor Williams
Okehampton, Devon

A The reason that freshly boiled water is more effective for making tea than water boiled twice is that the fresh water has a higher oxygen content. This should result in a tastier cuppa because more tea will be extracted from the tea leaves.

This can be easily demonstrated by placing a measured amount of tea leaves in two glass tumblers and adding freshly boiled water to one and repeatedly boiled water to the other. Examination of both tumblers after three minutes will reveal a much stronger brew from the freshly boiled water.

J. R. Stafford
Marks & Spencer, London

I was told, as a child, that the reason for using freshly drawn water to make tea was because the dissolved oxygen made the tea taste better. Water which has been standing or, worse, had been boiled, contained less dissolved oxygen. The British Standard 6008, which describes in great detail how to make a cup of tea, says that the water must be freshly boiling but does not say anything about it being freshly drawn. It also says that the milk should be put in the cup first to avoid scalding it.

As this British Standard is identical to International Standard ISO 3103, the supplementary question is why can't I get a decent cup of tea abroad?

N. C. Friswell
Horsham, West Sussex

The traditional explanation for making tea with freshly boiled water is that prolonged boiling drives off the dissolved oxygen, making the tea taste 'flat'. My own experiments with water simmered for an hour against freshly boiled water produced little perceptible difference, even though high-quality leaf tea was used and brewed for five minutes.

I would be surprised if the difference was of the slightest practical importance for tea made by dunking a tea bag, especially if the water had merely been boiled twice.
David Edge
Hatton, Derbyshire

I see that at least one reader remains unconvinced on the need to use freshly boiled water for tea.

Once, during an emergency overseas, we were instructed to boil all drinking water for several minutes. It didn't seem to affect the tea. However, we decided that it would be a good idea to use a domestic pressure cooker to raise the water temperature to beyond boiling point to sterilize the water thoroughly. This was fine when used for drinking or cooking, but when we tried using it for making tea the result was absolutely dreadful.

On the other hand, I have drunk tea at an altitude of 2100 metres where, of course, the boiling point is lower than 100 °c, but I noticed no difference in the taste. Nor did my tea-planter hosts make any comment on the point.

Pressure-cooked water apart, I think the length of time the tea is allowed to infuse is a more critical factor.
A. C. Rothney
East Grinstead, Surrey

Your correspondent A. C. Rothney may be surprised to hear that his/her shiny pressure cooker probably caused his/her nasty tea. Dissolved aluminium in the water, not the higher temperature to which the water had been subjected, is the reason the tea tasted awful. In the days when most kettles were made of aluminium they carried instructions to prepare the new kettle by repeatedly boiling fresh water and then discarding it. Only then should the first pot of tea be made with fresh water. During these repeated boilings, a patina of dull oxide built up inside the kettle and prevented the water dissolving the pure aluminium.

Lorna English
London

The preference for fresh water when making tea has little to do with oxygen but is related to dissolved metal salts (mainly calcium and magnesium bicarbonates, sulphates, and chlorides) which are present as impurities in tap water and affect the colour and taste of tea.

The effect of metal salts on the colour of tea can be demonstrated by comparing a brew made with freshly boiled pure water (deionized or melted freezer frost) with tea made with freshly boiled tapwater. The salts in tapwater give a darker brew, which is cloudier as a result of precipitated insoluble salts such as tannates.

Boiling tap water destabilizes the bicarbonates (so-called temporary hardness) which precipitate out as insoluble carbonates on cooling (this is why a kettle furs up with time). In hard-water areas, where more dissolved salts are present, repeated boiling and cooling will remove sufficient calcium and magnesium salts, although boiling for a long time without cooling has less effect.

There are three reasons why repeatedly boiled and cooled water can produce a less palatable tea. First, some of the precipitated carbonate remains in suspension, even after reboiling, as a white scum (particularly noticeable in new plastic kettles) and this taste is more marked than bicarbonates dissolved in water—especially when the scum interacts with the tea.

Secondly, the salts in the water which are not destabilized by boiling (so-called permanent water hardness) are gradually concentrated by evaporation, producing unpleasant flavours.

Finally, traces of metals, such as iron and copper, can accumulate in repeatedly boiled water and these can interact with oxygen and reducing agents in the tea (phenols) by complex redox reactions to produce further effects on flavour.

M. V. Wareing
Braintree, Essex

A As a caffeine addict, I suffer severe headaches if I go more than a day without my cups of tea. To conserve fuel on hikes lasting a number of days, I tried leaving a tea bag in a bottle of cold water for a few hours. It worked. Not only did it give me my fix of caffeine, but it tasted like tea, albeit cold tea. I haven't yet tried making such a cold infusion, then heating it in a microwave, but it should prove quite drinkable.

Syd Curtis
Hawthorne, Australia

A I read your reply to this question with dismay. The truth runs counter to A. C. Rothney's ideas.

My father was a tea taster and faultless at detecting whether we had boiled the water for too long. How did he do it?

Hard waters (and most waters do have some mineral salts in solution causing hardening) brew slower than soft or alkaline waters. If you boil hard water for considerably longer than the standard half a minute or so, more of the dissolved salts deposit themselves on the inside of the kettle. The emerging water is then softer than expected and softer than the tea taster balanced the tea blend for. It will brew quicker and with a darker colour than usual.

Tea manufacturers ensure constant performance by balancing their blend differently for sale in different water areas of the country, even where the brand label is the same. Hard water can be artificially softened with a pinch of bicarbonate of soda, but the dramatic darkening of the colour and change of flavour are unacceptable to most people—including tea tasters.
Bernard Howlett
Loughton, Essex

TWISTER

Q ***Here in Zimbabwe we buy milk in plastic packets. Most people cut a tiny piece off the edge of the packet to pour the milk. I have noticed that when the milk leaves the packet under pressure it exits in a corkscrew or a spiral fashion. Of course, other liquids would behave in a similar fashion. What forces operate to allow an unchannelled liquid to follow this path? I have noticed that the smaller the opening in the packet the greater the amount of twisting in the path followed by the milk.*** 
David White
Chinhoyi, Zimbabwe

A The corkscrew effect you notice is just the bottom end of the whirlpool that is occurring inside the carton as the milk exits. The force that causes it is usually called the Coriolis force. This is responsible for all whirlpooly stuff you might find. Milk cartons and bottles give you the same effect, but it is less noticeable because of the shape of the cross-section of their openings.

When the milk leaves the contents under pressure from squeezing the sachet while pouring, you effectively increase the liquid's speed. This increases the Coriolis force—which is proportional to the speed of an object in a rotating inertial frame, as well as the frame's angular velocity and the distance from the object to the axis of rotation. This gives a tighter corkscrew. In effect, milk under pressure screws up.

John Lenton
Cordoba, Argentina

The twisting of the stream of milk coming out of the package has more to do with the shape of the hole (usually a long, thin one), the difference in pressure on the milk from one side of the exit hole to the other, and the force of the surface tension between the milk and the side of the container. It has nothing to do with the Coriolis force as suggested by your correspondent.

The Coriolis force is a real phenomenon. Because the Earth rotates, a fluid that flows along the Earth's surface feels a Coriolis acceleration perpendicular to its velocity. In the northern hemisphere, Coriolis acceleration makes low-pressure storm systems (hurricanes) spin anticlockwise. But in the southern hemisphere storm systems (typhoons) spin clockwise because the direction of the Coriolis acceleration is reversed.

This large-scale meteorological effect leads to the speculation that the small-scale bathtub vortex that you see when you pull the plug from the drain spins one way north of the equator and the other way south of the equator. This is incorrect, the Coriolis force is far too small to have an effect on the direction of bathtub whirlpools or twisting milk coming out of cartons.

The force can be seen in a tub of water only

under controlled experimental conditions including a symmetrical low-friction tub, tight control of thermal currents, and letting the water stand for a long time (a day or more) so the residual fluid motion from filling has ceased.

Raymond Hall
No address supplied

A Your first correspondent's answer to the question is not entirely correct. While it is correct to say that the corkscrew is the end of the whirlpool occurring within the packet, he is wrong to suggest that the cause of the whirlpool is the Coriolis effect.

Instead, the 'ice-skater' effect is responsible. Any small wobbles you have given the milk packet will set the fluid moving inside in one direction or another. As the fluid moves out through the small hole, its angular momentum is conserved. That means that, as it moves into a smaller diameter stream, it spins faster—just as ice-skaters spin faster when they pull their arms in closer. This is also why the corkscrew effect is enhanced with smaller openings.

Sonya Legg
California

NOW WASH

Q ***Do normal hand soaps actually kill germs? We are always encouraged to wash our hands after a visit to the toilet, but does this really make much difference? I remember a biology experiment at school where we pressed our fingers into a Petri dish and cultured the bacteria that were deposited***

there. As I remember there was no difference between the two.

Philip Teale
No address supplied

Ordinary soap has little value as an antiseptic and it does not kill or inhibit many types of bacteria. However, the important function of washing one's hands with soap and water is the mechanical removal of bacteria through scrubbing.

The skin normally contains dead cells, dried sweat, bacteria, oily secretions, and dust. Soap emulsifies this mixture and the water washes it away.

Because soaps are good for removing bacteria from the skin it is important to teach children always to wash their hands before eating and after going to the toilet as this prevents the transfer of potentially harmful bacteria by the faecal to oral route. Many cosmetic soaps contain antimicrobial compounds that strongly inhibit Gram-positive bacteria and these are used to decrease body odour by preventing microbial growth on body secretions.

There are non-household soaps that do contain some type of antibacterial compound. These compounds act as disinfectants and kill bacteria. Of course, during our everyday activities we do not need or want completely to disinfect or sterilize our skin because the normal population of bacteria on us acts as a barrier against infection by pathogens.

Graham O'Hara
Murdoch University, Australia

No, ordinary hand soap does not kill germs. In fact, many common bacteria will multiply quickly on a wet bar of soap and you could finish up with more on your hands after washing than before. So try to keep your soap dry.

But although soap doesn't kill the bugs, the action of washing does two things. Having been to the toilet you will probably have contaminated your hands with faecal bacteria, no matter how many times you folded the paper.

The first function of washing is the physical removal of the vast majority of these bugs. The soap makes sure that the water is wet enough to free the bugs from your skin, and a good scrub with a nailbrush can remove as much as 99 per cent of the microbes.

The second function is, however, much more important. Healthy skin is a bacterial battleground populated with 'friendly' bugs. These eat our sweat and, in return, defend our skin from less friendly bugs that would not only eat our sweat but us as well. *Staphylococcus aureus* is a typical invader that causes pimples and boils (or worse) when it beats our defences. When we wash, we not only remove dirt and invasive bacteria, we also release a tide of friendly bugs from pores to recoat our skin for protection.

This is fine unless you work in food preparation; you will then be required to use soaps that contain broad spectrum bactericides that kill all types of bacteria indiscriminately.

This is all well and good for food hygiene but, when your front-line guards are destroyed, watch out for skin invaders when you step back into the real world, where bacterial warfare reigns supreme.

Derek Smith
No address supplied

Effective hand washing depends on many factors—the soap or other cleaning agent used, clean running water, clean cloth if used, clean towel or air drier for drying, and good technique.

Normal hand soap can remove germs, if used properly, and left to drain and dry between uses, not left sitting in a puddle of soggy soap and stagnant water. Pump dispensers are generally better than soap bars. Water must be clean and taps must not be contaminated by dirty hands—that's why hospital sink taps have elbow levers or, more rarely, foot pedals.

Effective rinsing and drying, to remove any contaminated water without adding further contamination from a damp towel, are vital components of proper hand washing technique. This is why disposable paper towels are usually used in hospitals.

Studies on hand-washing techniques by nurses have shown that some areas of the hand are less well cleaned than others—fingers and the web between thumb and first finger are commonly inadequately cleaned.

Jean Sinclair
Solihull, West Midlands

The reason why you may not have been able to differentiate between the cultures of bacteria from washed and unwashed hands could be because of the way in which bacteria grow under conditions of limited resources. A graph of population against time shows a sigmoidal shape. Initially, the population increases in an exponential manner and then levels out as food and space begin to run out. There may have been more bacteria from the unwashed hand but this could not be determined because both populations eventually reached the limiting maximum and appeared identical.

Adnaan Ali
No address supplied

A When I had a wart removed from my hand a few years ago my doctor told me an interesting fact about the microbial risks of washing your hands. There is a variety of wart which specializes in colonizing wounds and can be recognized because it is found in lines on skin that have been caused by grazes. The doctor told me it is common in surgeons because they scrub up before operating.
David Frin
London

Those crying eyes

Q ***Most of us suffer from watering eyes when we chop onions. One surprising solution to this problem is to trickle cold tap water onto your wrists for a few seconds as soon as you feel any discomfort. Try it, it works; but why?***
Yanni Papastovrou

A It is well established that the route taken by the lacrimatory material to the tear glands is through the nose. It is part of folklore that nosebleeds resulting from rupture of tiny vessels in the mucous membrane can often be arrested by a sudden chill. Hence the remedy of putting a cold bunch of keys down the patient's back. There is every likelihood that this vasoconstriction effect tends to block the movement not only of blood but also of any lacrimator, such as the smell of onions.

Readers might find it inconvenient to peel onions with water running down their arms so a peg on the nose, or two little bungs made of kitchen roll suggest themselves as an alternative.
E. Barraclough
Cornwall

A Your correspondent should peel and cut onions from the top, leaving the root end untouched until the final cut. There is no need for such things as wet wrists or matches in the mouth.
Susan Hillyard
Bakewell, Derbyshire

A To prevent watering eyes when slicing an onion, try putting the onion in your freezer about half an hour before slicing.
David Krupp
No address supplied

A A more practical method to avoid crying when chopping onions is first to slice it with transverse cuts, starting at the top and working towards the bottom.

Be careful not to cut into or through the root, though. Next make a second series of parallel cuts from the onion's top to its bottom, at right angles to the first set of cuts. Remove the root last.
Fred Startz
Jakarta, Indonesia

LIGHT ALE

Q ***A strange phenomenon caused much comment at a barbecue this summer. To keep the cans of beer cool they were placed in a large plastic dustbin that was filled with cold water. Strangely, some cans remained on the bottom of the bin while others floated on the surface of the water. There were more submerged cans than floating cans, but this behaviour was not confined to any particular brand. What could be the reason for this?***
Charlotte Harding Rains
Sawbridgeworth, Hertfordshire

This owes a little to the laws of hydrostatics and a lot to consumer laws. The volume of beer that is printed on the side of a can is a legal minimum, and so beer cans are designed to accommodate a certain amount of extra liquid.

The exact amount of extra beer contained in the can depends on the brewer's quality control procedures during can filling. If the beer is accurately metered during filling, then cans need to be only slightly larger than the nominal volume of beer they contain. Where the filling process is less accurate cans must be even larger, cost more to make and transport, and may be filled with more beer for which the brewer cannot charge.

When the can is completely filled with more than the quantity of beer that is printed on the side it has a net density that is more than that of water (beer and aluminium being denser than water). When the can is filled with the minimum quantity of drink marked on the side, the rest of the can is filled with air and carbon dioxide, making the overall density less than when it is full.

Assuming that beer and air have densities of 1010 kilograms per cubic metre and 1 kilogram per cubic metre respectively, and that a nominal 440 millilitre aluminium can has a mass of 30 grams, a few rough calculations (based on the cans in my fridge) would suggest that around 8 per cent extra volume of air on top of the standard 440 millilitres of beer would be enough to make it float. This can be measured by the fill-level of the can: in this case it would be about 9 millimetres from the top.

The fact that there were more submerged than floating cans would confirm that brewers do over-fill cans to comply with weights and measures regulations.

So, if your thirst needs plenty of quenching,

or you just want more beer for your money, you should go for the cans that don't float.
Phil Kerry
Buxton, Derbyshire

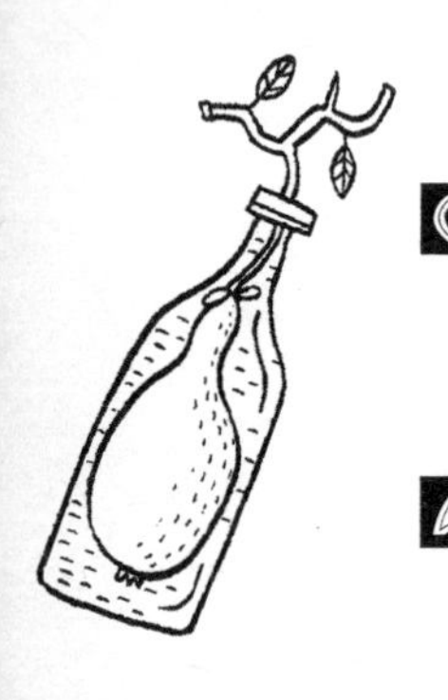

BOTTLED UP

Q ***How do they get a fully grown pear inside a bottle of Poire William liqueur?***
Jonathan Barnett
No address supplied

A The bottlers go into the pear groves and place empty bottles over the pear buds. As they mature, they grow in the bottle. When nearly ripe, they are picked and then the bottles are filled with the liqueur.
Steve Andrews
New York

A I have successfully grown pears inside bottles a number of times. Wait until the blossom has dropped off and the pear is at its earliest stage of development. Place the bottle over this proto-pear, tie it securely in place, and allow the fruit to mature inside the bottle as normal.

I find that it helps to blow air through the bottle (or to shade it) on particularly hot days, and to remove any competing pears from the same branch. This helps to stop the fruit rotting, and directs all the nutrients to the pear growing inside the bottle. It can be cut from the branch when it is mature.

Refinements, such as peeling the pear inside the bottle, call for a good set of appropriate tools and a steady hand.
Gavin Whittaker
Heriot, Midlothian

Some Japanese farmers grow melons inside cube-shaped containers. As they grow they fill the container and end up as cubes. Although edible, these melons lack flavour and are primarily ornamental.
Edward Edmondson
Barnwood, Gloucestershire

The idea of bottling fruit while it is still on the tree has been taken up in California, where they grow aubergines in plastic moulds to make them look like the faces of celebrities such as Elvis and Bill Clinton. Always expect California to go one better.
Henrik Schmidt
Copenhagen, Denmark

Kettle hums

Why does a kettle sing? Why does the note rise at first, then fade for a while, and then return with a falling frequency?
Don Munro
University of Newcastle, Australia

If you leave the lid off your electric kettle and switch on, you can see what is happening. The heating element quickly becomes covered with small silvery bubbles, each about 1 millimetre in diameter. These are air bubbles, forced out of solution by heat from the element. Rough parts of the element's metal surface provide nuclei for their growth and they eventually detach from the hot element and rise to the surface. These bubbles form and burst silently, and are clearly not the cause of the kettle singing.

After about a minute, the air bubbles are

replaced by innumerable smaller bubbles of superheated steam that cling to the growth nuclei on the heating element.

A few seconds later, these primary steam bubbles become unstable. As each bubble forms, its buoyancy tends to pull it away from the hot surface. Being surrounded by water which is still far below boiling point, the primary steam bubble suddenly condenses, collapsing implosively. Curiously, the bubble does not vanish completely, but leaves behind a minute secondary bubble, presumably of water vapour, that does not immediately condense but is whirled away by the convection currents. Soon there is such a cloud of these secondary bubbles that the water becomes turbid for half a minute or so.

Meanwhile, the shock waves transmitted through the water by the imploding primary bubbles produce a sizzling sound. You can give this sound a more definite pitch by temporarily replacing the kettle lid. This defines a volume of air above the water surface that resonates to some of the frequencies present in the shock waves.

Soon, the cloud of secondary bubbles clears, and there is a general increase in the size of the primary steam bubbles that are still forming on the element. These are no longer forced to collapse immediately and implosively, because the surrounding water is now practically at boiling point, so the noise fades away. As they grow, streams of buoyant primary bubbles detach themselves from the surface of the element, condensing in the cooler water a centimetre or so above it.

Within seconds, the water becomes hot enough to allow large detached primary bubbles to reach the surface, and now you can hear a return of

sound caused by the low gurgle of their bursting in the air cavity above the water.
Roger Kersey
Nutley, East Sussex

GADGETS AND INVENTIONS

swag

KEEP HIM TALKING

Q ***How, in the days before digital exchanges, were telephone calls traced? And why did it always take so long to do (at least in films, where the villain always hung up just before the trace was made)?***

David Prossner
Oxford

A Twenty or more years ago, as a young trainee, I was coached in the covert skills of tracing. The principle was simple but relied on fleetness of foot. The technology of the day was Strowger switching, which consisted of electromechanical devices that connected calls through a two-dimensional—horizontal and vertical—matrix of electrical contacts. Each contact was connected to another switch—and so on until it reached the called telephone. My task was to inspect the first switch, see which contact the wiper blades were resting on, run to the next switch, inspect that one, and so on. This process could involve running up and down stairs in the chase to trace the call before the handset was replaced.

Bryan Parish
Ubley, Somerset

A The familiar TV depiction of 'tracing' calls should more accurately be described as a 'back trace'. To explain the process it is necessary to understand the way in which a call was set up in the days before electronic exchanges.

Upon lifting the handset, a signal was sent to a piece of equipment called a uniselector. This rotated over a semicircular set of contacts to find a free switch called a first selector, which was a two-motion device. As its name implies, this selector could move both vertically and horizontally. The dial tone was returned from this selector, and the selector was stepped vertically under control of the dial.

Horizontal, or rotary, motion then took place automatically to find a second selector, where again vertical movement was under control of the dial and rotary action took place automatically to find a final selector. The final selector then accepted the third digit to give vertical motion, and the fourth digit to give rotary motion. Thus, if the number dialled was 4567, the first selector would accept the 4, the second selector would accept the 5, and the final selector would accept the 6 and 7 to call the required number.

A forward trace was a relatively simple task. All that was required was to follow the call from switch to switch using the grading charts, which gave information as to where connections were made. Bearing in mind that a uniselector had 50 outlets, any one of which could be selected to find a free first selector, the first selector had 20 outlets to find a free second selector, the second selector also had 20 outlets to find a free final selector, and any one of up to 300 final selectors had access to the called number, it could take up to 15 minutes to do a forward trace.

To trace a call from a called number back to the calling number is a bit like only knowing the meaning of a word and then trying to find the word in a dictionary. A great deal of searching over 3-metre-high racks of equipment in narrow rows, sometimes 45 metres long, was required. You had to use ladders, and in larger exchanges calls were traced on different floors.

This procedure applied only if the call was from and to the same exchange. If the caller made a call

to another exchange the process had then to be repeated at the final destination exchange.
Steve Lyne
Norwich

STOP-GO

Q ***Why are the coloured lights in traffic signals universally arranged red over amber over green, as opposed to the universal practice of railway signals which have green over amber over red (for a three-aspect signal)?***
Roger Henry
Parkes, Australia

A The difference between road and rail usage derives from the history of railways and the primacy of safety. The old mechanical railway signalling arms were designed so that failure, which would be in the 'down' position, meant stop. The illuminated part of the signal consisted of two coloured glass panels in the far end of the signal arm, beyond the pivot, which moved in front of a fixed lantern. Even though the higher of the two glass panels was the red panel, it showed when the signal was down and this meant stop. While railways retained mixed mechanical and electrical signalling the signals had to be compatible. Therefore, the new electrical signals showed red at the bottom so that train drivers always equated either signal in its down or bottom position with the order to stop.

Road signals had no mechanical forerunner and are designed so the most important light, the red, can be seen from the greatest distance. This means putting it as high as possible. Anyone who

has used Junction 3 of the northbound M25 at night will appreciate this. Additionally, visibility for railway signals is not the same issue it is on the roads. Railway signal sites are carefully selected.
Gerald Dorey
Oxford

Gerald Dorey is only partly correct in his historical explanation for the order of railway signal lights. Indeed, he overlooks the large parts of the country where lower quadrant semaphore signals (in which horizontal means danger and 45° down means clear) were used. In these signals the red light was therefore at the top.

The main reason for the red light being at the bottom in modern British signalling installations is the weather. To ensure visibility in bright sunshine, each colour light has a long cowl or hood above it. In the winter, however, snow can build up on these cowls and obscure the light above. Being at the bottom, the most safety critical red light has no other light below it and therefore no cowl, so snow cannot build up to cover a red light.
Vincent Luthart
London

There are two kinds of mechanical, or semaphore, signals. In the older lower quadrant type, the arm slopes downwards from the pivot to show clear or green, is returned to the horizontal by a counterweight, and the lamp glasses are red above green. In the newer upper quadrant type, the arm slopes up for clear, returns by its own weight (as in the scene in the classic film *The Lady Killers*) and the lamp glasses are in fact side-by-side. Red is nearer the pivot and green is to its right on the outside.

In both, the horizontal arm means stop but this is not synonymous with down, which means opposite things in the two cases. Red arms are always used in stop signals but distants (meaning warning) operate similarly. However, on distants the arm and lamp glass are yellow, not red, and these mean pass with caution.

The arrangement of multiple-aspect, coloured light signals has nothing to do with that of arm position. The red is at the bottom simply because it is the position nearest the driver's eyes; yellow is above, then green, and, in four-aspect signals, the second yellow is top-most, above the green.

C. C. Thornburn
Aston University, Birmingham

Road users do not have to pass a colour vision test, and therefore the position of the red, amber, and green lights must always be the same, so the light illuminated can be recognized by position as well as colour. Such signals are usually placed at sites where a speed limit applies, and, because of the higher braking coefficient of rubber tyres, the driver can still stop safely even after identifying a red indication only by its position.

The train driver, whose colour vision is checked regularly, has to act on signals at a far greater distance to ensure the train can be stopped in time. On main lines the indication has to be identified accurately at long range, when it is impossible for the driver to see its position and has to rely solely on its colour.

The original question was, in fact, incorrect, because there is no universal railway layout of green over yellow over red (in railway parlance, the caution indication is referred to as yellow, not amber). In the past, some signals only had a single lens, the different indications being given by interposing coloured filters over the beam. The only fixed rule with the layout of signals with multiple lenses is that the one which exhibits the red indication is mounted nearest to the line of the driver's eyes. In some places, therefore, it may be at the top, as it is with a road traffic signal.

On high-speed lines, it is necessary to have a double-yellow indication, which is exhibited by the signal before the one showing a single yellow for caution. That, in turn, will be a further three-quarters of a mile or so before the one showing red for stop. This gives a run of two signals warning of a stop signal ahead. Such double-yellow signals normally have the two yellow lights separated by the green one, to maximize their visual separation when viewed from a distance.

P. W. B. Semmens
York

STICK WITH IT

Q ***Why doesn't superglue stick to the inside of its tube?***

Ajit Vesudevan
Oxford

A Superglue will not stick to the inside of its tube because the tube contains oxygen in the form of air but excludes water. Oxygen inhibits whereas water catalyses.

Yvonne Adam
Bostik Limited, Leicester

A Superglue doesn't stick to the inside of the tube because, being based on a cyano-acrylate monomer, it requires moisture in the form of water

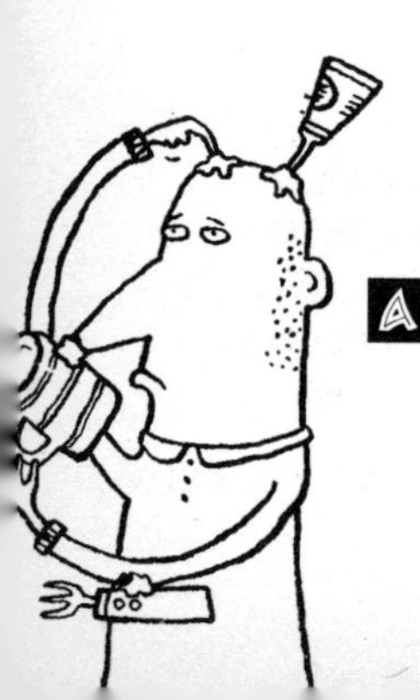

or some other active hydrogen-bearing compound to polymerize.

This explains why the best join between two surfaces is made using a thin glue line. An excess thickness of glue will lead to a retarded cure. This moisture sensitivity explains two things. First, why the bottle comes with a seal that's impossible to break without covering oneself in glue and why the resulting spillage adheres so well to your skin—being warm and moist, skin makes an ideal substrate.

Brian Goodliffe
Wetherby, West Yorkshire

A The Loctite company in the US discovered the inhibition by oxygen of the otherwise rapid polymerization of cyano-acrylate. That is why the bottle must always be left with plenty of air inside. The liquid monomer converts to solid polymer when oxygen is excluded by trapping it between close-fitting surfaces.

E. Barraclough
Otterham, North Cornwall

Controlled blast

Q ***Why can I cancel out the hissing noise of my gas fire by aiming an old-style ultrasonic TV remote control at it?***

Stephen Marsden
Draycott, Derbyshire

A The effects of sounds on naked gas flames appears to have been first noticed by Professor Leconte at a musical party in the US in 1858. Professor Tyndale gave an evening discourse on sound and sensitive flames at the Royal

Institution in 1867. This ended with a gas flame curtsying to the notes of a music box.

In 1966, Frank Briffa and I observed that a small amount of ultrasonic energy directed across an orifice would reduce the noise from a gas jet, lit or unlit (this now forms British Patent 1 147 103). This was demonstrated in the library of the Royal Institution a century after Tyndale's presentation.

Acoustic waves that are directed across the top of a jet orifice cause disturbances in the gas flow which grow to form vortices. These are shed downstream from the nozzle on either side of the jet. This gives a broader and more stable flame, which emits less noise. Only a small amount of ultrasonic energy is required, a transducer using less than 0.01 per cent of the energy produced by the flame gave a noise reduction of 5 dB(A). An account with schlieren photographs of the gas jets with and without the ultrasonic field was published in *New Scientist* ('More sound means a quieter flame', 18 June 1970).

Although our work at the Egham Laboratories of Shell Research was mainly involved with reducing noise from large industrial burners, there was general concern at the time that domestic gas fires being converted from town gas to natural gas were much noisier after the conversion. We found that this could be eliminated by the combination of a small ultrasonic source and a tapering of the gas nozzles. However, people soon got used to the extra noise of natural gas.

At one time there was even press speculation that ultrasonics could be applied to aircraft engines. So do not ditch your old ultrasonic remote controls.

Roy Fursey

Netherbury, Dorset

PRESSURE SITUATION

Q ***We are all familiar with the popping ears associated with take-off and landing in an aeroplane. This is caused by changes in pressure but, because the aircraft cabin is artificially pressurized, why isn't the internal pressure maintained at one level throughout the journey?***

Craig Lindsay
Aberdeen

A For reasons of fuel economy, large civil transport aircraft have to fly at altitudes far in excess of those capable of sustaining life. Whereas 5500 metres is about the maximum altitude at which a person can live for any extended period, a subsonic passenger jet has the best fuel economy when flying at around 12 000 metres.

Aircraft manufacturers, therefore, have no choice but to pressurise the interior of a passenger aircraft. This poses huge technical problems. At 12 000 metres, where the pressure is about one-fifth of that at sea level, the pressure inside is trying to burst the fuselage apart. This pressure has to be contained and all the stretching and flexing of the fuselage during a flight has to be kept within safe limits. It is far easier to do this—if the pressure differential between inside and outside is kept to a minimum, a cheaper and lighter fuselage structure can be used.

For civil airliners this means that the pressure inside during cruising is kept at the lowest possible safe level—2500 metres. This is about the maximum altitude which a normal healthy person can be subjected to without ill effects.

Even so, unfit people, those with respiratory illnesses, and those who have sampled a few too many duty-free drinks might still feel ill, even at this altitude.

There is another problem: all airfields are not at the same altitude. In an extreme case, a flight from Heathrow in England to La Paz in Bolivia would entail going from sea level to around 5200 metres, where the air pressure is about half that at sea level. Under these circumstances it is just not possible to maintain the same pressure throughout the flight. Imagine what would happen if the pressures inside and outside were not the same at the time the doors were opened: the effect would be quite spectacular and most undesirable.

As for the ear popping: nowadays, 'for your safety and comfort', the internal pressure is imperceptibly reduced, all under computer control, as the aircraft climbs. It is gradually increased (or, in the case of La Paz and other high altitude airports, decreased) during descent so that, as the aircraft is coming to a stop on the runway, the pressure inside and out is the same. This is normally sufficient for your ears to adjust, but if all else fails, pinch your nose and gently but firmly increase the pressure in the nasal cavity until you feel the pressure equalize.

Terence Hollingworth
Blagnac, France

An advantage of flying by Concorde is that the fuselage has to be especially strong to fly at very high altitudes, so the cabin pressure does not have to be reduced below that experienced at 900 metres.

Arthur Cox
Alton, Hampshire

TROUBLESOME TIN

Q ***Why does corned beef come in that uniquely shaped tin which is impossible to open when the key is missing (as it often is)?***

Rob Davies
Anstey, Leicestershire

A My father is something of a corned beef connoisseur and, consequently, whenever I go abroad I am always on the lookout for new and exciting brands for him to try. I have collected samples from many countries in Europe, Oceania, and the Americas. The tapering rectangular tin-with-key variety seems to be fairly ubiquitous. However, I did come across a local Papuan brand which came in a larger version of the more familiar flat cylindrical tin, typical of salmon or tuna.

Unlike salmon, which is best enjoyed in flakes, corned beef is best chilled so that it can be removed from the tin in one piece and then be sliced for sandwiches. However, opening the cylindrical Papuan tin with a standard tin-opener makes extracting the product whole almost impossible.

Even removing both the top and bottom of the Papuan tin left a rim which gouged the meat when pushed through, resulting in a far messier procedure than that facilitated by the rectangular tin. The key on a rectangular tin peels away a precut strip, separating the top and bottom of the tin and leaving an opening with no rims which allows the product to slide out whole and unscathed.

The best tins place the removable strip close to the base, and taper towards the top. Squeezing the base near the peeled strip grips the product, and the taper allows the top of the tin to slide away quite easily, leaving the delectable morsel exposed on the merest scrap of remaining tin.

Argentinian corned beef, the clear winner for taste and consistency according to my father, also comes with the most firmly attached keys.

Alex Russell
Edmonton, Canada

A This is a good example of a design that is not fail-safe. If you have the key, the tapered tin allows the meat to be extracted more easily, but if the key is missing, you have a problem unless the tin has a rim which will engage with a standard tin-opener. The solution is to save a few keys. However, the strip of tin which remains on the key after use must be unwound very carefully because it has sharp edges.

J. M. Woodgate
Rayleigh, Essex

Sticking doors

Q ***My fridge door is kept shut by a magnetic strip in a flexible plastic seal. If I attempt to open the door soon after closing it, it seems to be held shut by a vacuum for a few seconds: why is this?***

Christopher Nunn
East Grinstead, West Sussex

A Fridge doors stick in this way because opening the door allows some of the cold air to flow out of the bottom of the fridge. If you stand in bare feet at the door of an open fridge you will feel this cold, dense air. This allows warmer air at room temperature into the top of the fridge.

When the fridge is closed again, this new air cools and contracts, creating a partial vacuum and making the door seem to stick. The effect is most noticeable with a freezer compartment, because the lower temperature of the freezer creates an even greater partial vacuum than a fridge does when air at room temperature is introduced, as the air undergoes even more contraction.

I have often wondered if the freezing of the water vapour contained in this introduced air adds to the effect inside the freezer compartment. Notice how ice gradually builds up in the compartment of non-defrosting freezers.

To open the freezer door when the temperature contrast between room temperature and freezer temperature is large (on, say, a particularly hot day), it is sometimes necessary to prise open the plastic seal a little to allow some air in to equalize the partial vacuum in the freezer compartment.

The partial vacuum created by opening the door soon passes, however, because the door seals are not airtight and the air pressures inside and outside the fridge are equalized by outside air leaking into the fridge compartments.

This means that if you had no idea what the original question was referring to, you should check the quality of the seals on your fridge door—they may be leaking a lot of cold air and consequently making the fridge work harder than it should have to.

Paul Bishop
Monash University, Australia

BAGS OF AIR

Q ***On a recent plane flight, I started wondering what purpose the small bag serves on the emergency air line. The bag is between the air supply and the emergency mask that drops if there is a sudden loss of cabin pressure. What is its function and why doesn't it inflate?***

Ean Warren
California

A The function of the bag is to act as a reservoir of oxygen. During some phases of respiration the flow into the lungs can be more than 30 litres per minute, yet oxygen isn't provided from the supply at this high rate.

So, during those phases of respiration when the flow into the lungs is less than the flow coming from the oxygen supply, and during breathing pauses, oxygen enters the bag and fills it up. Then, during periods when the higher intake is needed, the passenger can inhale the oxygen stored in the bag. Reservoir bags avoid the need for wasteful high flow rates of oxygen.

A. Speakman
Wigan, Greater Manchester

GUNG-HO GUNS

Q ***In many parts of the world, people celebrate victories, birthdays, and similar events by firing guns into the air with great exuberance and a seeming disregard for the welfare of themselves and others.***

Assuming the barrel of the gun is perpendicular to the ground when the bullet leaves it, approximately what altitude would it reach and what is its velocity (and potential lethality) when it falls back to Earth?

Leo Kelly
Auckland, New Zealand

Firing handguns into the air is commonplace in some parts of the world and causes injuries with a disproportionate number of fatalities. For a typical modern 7.62 millimetre calibre bullet fired vertically into the air from a rifle, the bullet will have a velocity of about 840 metres per second as it leaves the muzzle of the gun and will reach a height of about 2400 metres in some 17 seconds. It will then take another 40 seconds or so to return to the ground, usually at a relatively low speed which approximates to the terminal velocity. This part of the bullet's trajectory will normally be flown base first because the bullet is actually more stable in rearward than in forward flight.

Even with a truly vertical launch, the bullet can move some distance sideways. It will spend about 8 seconds at between 2300 and 2400 metres and at a vertical velocity of less than 40 metres per second. In this time it is particularly susceptible to lateral movement by the wind. It will return to the ground at a speed of some 70 metres per second.

This sounds quite low but, because of the predominance of cranial injuries, the number of deaths and serious injury as a proportion of the number of gunshot wounds is surprisingly high. It is typically some five times more than is observed in normal firing. As might be expected, measurements of this kind are rather difficult and the above values come from a computer model of the bullet flight.

Sam Ellis and Gerry Moss
Royal Military College of Science
Swindon, Wiltshire

Different bullet types behave in different ways. A .22LR bullet reaches a maximum altitude of 1179 metres and a terminal velocity of either 60 metres per second or 43 metres per second depending upon whether the bullet falls base first or tumbles.

A .44 magnum bullet will reach an altitude of 1377 metres and will have a terminal velocity of 76 metres per second falling base first. A .30-06 bullet will reach an altitude of 3080 metres with a terminal velocity of 99 metres per second.

The total flight time for the .22LR is between 30 and 36 seconds, while for the .30-06 it is about 58 seconds. The velocities of the different bullets as they leave the rifle muzzle are much higher than their falling velocities. A .22LR has a muzzle velocity of 383 metres per second and the .30-06 has a muzzle velocity of 823 metres per second.

According to tests undertaken by Browning at the beginning of the century and recently by L. C. Haag, the bullet velocity required for skin penetration is between 45 and 60 metres per second, which is within the velocity range of falling bullets. Of course, skin penetration is not required in order to cause serious or fatal injury and any responsible person will never fire bullets into the air in this manner.

The questioner may like to read 'Falling bullets: terminal velocities and penetration studies', by

L. C. Haag, Wound Ballistics Conference, April 1994, Sacramento, California.
David Maddison
Melbourne, Australia

John W. Hicks in his book *The Theory of the Rifle and Rifle Shooting* describes experiments made in 1909 by a Major Hardcastle who fired .303 rifle rounds vertically into the air on the River Stour at Manningtree. His boatman, probably a theorist unaware of the winds aloft, insisted on wearing a copy of *Kelly's Directory* on his head.

However, none of the bullets landed within 100 yards, some landed up to a 400 metres away, and others were lost altogether.

Julian S. Hatcher records a similar experiment in Florida immediately after the First World War. A .30 calibre machine gun was set up on a 3-metre square stage in a sea inlet where the water was very calm so that the returning bullets could be seen to splash down. A sheet of armour above the stage protected the experimenters. The gun was then adjusted to centre the groups of returning bullets onto the stage.

Of more than 500 bullets fired into the air, only four hit the stage at the end of their return journey. The bullets fired in each burst fell in groups about 20 metres across.

The bullets rose to approximately 2750 metres before falling back. With a total flight time of about a minute, the wind exerts a noticeable effect on the return point.
Dick Fillery
London

In my youth, I used to collect brass cartridge cases ejected from aircraft machine guns during the Battle of Britain for salvage. They drifted down slowly from the sky because, I guess, their mass to surface area ratio was low. However, they were still warm when I picked them up.

Accordingly, if the projectile is small, like a .303 bullet, it does nobody much harm when it lands. Like a mouse in a mine shaft, its terminal velocity is negligible. However, if because of its mass the projectile has enough terminal velocity, it could kill you.

M. W. Evans
Inzievar, Fife

PORTHOLE PARADOX

Q ***Why, and since when, have the windows of a ship's hull been round?***

Campbell Munro
Oban, Strathclyde

A I assume that your correspondent is referring to old pictures and prints of wooden ships, where the portholes (usually gun ports) are square or rectangular, and is wondering why such ports are round in steel-hulled ships.

When ships were made of wood, the architectural material was fibrous and fairly flexible (wooden ships really did creak and this was caused by timber flexure from wave action). However, wood—especially wet wood—is highly resistant to fatigue stress. Try breaking a piece of wet willow by repeated reverse flexure, and then try the same process with a mild steel bar or rod of similar section. Ferrous metals (indeed, most metals), are highly prone to crystalline fracture as a result of changes to the grain structure arising from repeated stress reversal. The effect depends upon the section, heat treatment, carbon content, and any alloying elements present.

Towards the end of the nineteenth century, steel hulls became universal for merchant vessels, and subsequently for warships. Naval architects found out pretty quickly that any rectangular or square hole in a ship, whether on a deck (a hatch) or in the hull (a porthole) was a source of metal fatigue, commencing at the corners. The hull or deck would literally rip, due to flexure cycles brought about by wave action; the rougher the seas, the greater the magnitude of the stress. The unlucky sailors found that their ship was most likely to fall apart when weather conditions were at their worst. Thus, naval architects specified circular portholes, and radiused corners for deck hatches. This left no sharp corners for stress concentration.

David Lord
Aldershot, Hampshire

BORING QUESTION

Q **Why does making a bullet spin by 'rifling' the bore of a rifle barrel make it travel with greater accuracy?**

Philip Forward
Manchester

A Modern rifles use small calibre bullets. If they were spherical, their low mass per unit cross-section would mean that air resistance slowed them down excessively. This would shorten the range of the weapon and reduce its hitting power.

Therefore, for small calibre bullets to meet military requirements for range and flatness of trajectory, their mass had to be increased and a more streamlined form adopted. This meant that the bullet had to be made much longer and with a tapered nose.

Such a bullet will be inherently unstable if projected from a smooth bore. Any minute distortion or roughness of its surface or turbulence of the air will cause the axis of the bullet to deviate from the direction of projection. This deviation will increase rapidly until the bullet topples end-over-end. The result is gross inaccuracy and reduced range caused by the extra drag.

The bullet will only fly true if it spins about its axis with enough angular momentum to overcome these in-flight perturbations. In effect, the bullet becomes a gyroscope travelling along its axis of rotation.

However, it will still not travel in a straight line. Besides the parabolic trajectory caused by the fall of the bullet due to gravity, there is a second effect as the surface drag, accentuated by the grooving caused by the rifling, will cause the bullet to drift gradually in an increasing spiral.

The effects of inadequate spin can be quite spectacular. Towards the end of the Second World War, I took part in trials to assess the rifling life of a machine gun by firing some thousands of rounds continuously. We could see the barrel was worn out when the bullets started going sideways through a target placed only about 200 metres away.

Reginald Titt
Salisbury, Wiltshire

A spinning object has angular momentum and this is a conserved quantity. The direction of the axis of the rotation cannot change without changing the angular momentum. This means that a spinning bullet cannot change its direction easily—obviously very important to gun designers

and users. This is the same effect as in a gyroscope and it means that the bullet will point in the same direction throughout its flight which, ignoring the effect of gravity, makes it follow a straighter path.
David Feldman
Cockfosters, Hertfordshire

A In addition to accuracy, international law requires bullets to spin. Early American M-16 assault rifles from the Vietnam era had inadequately rifled barrels, which didn't give the bullets enough spin. This made the bullets tumble into the human target, provoking accusations that the Americans were using dumdum bullets, which cause greater injury than conventional ones.
Pekka Haussalo
Espoo, Finland

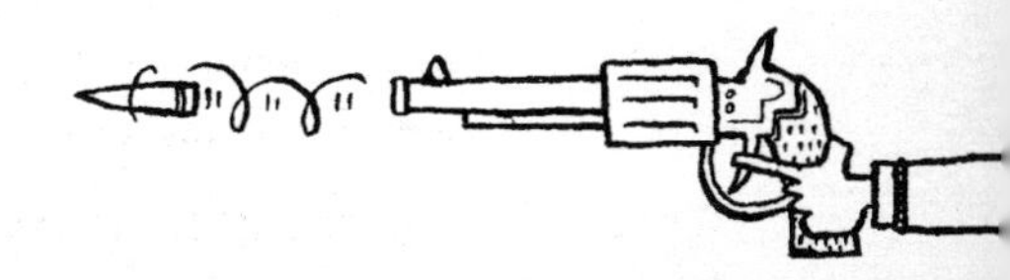

BUBBLES, LIQUIDS, AND ICE

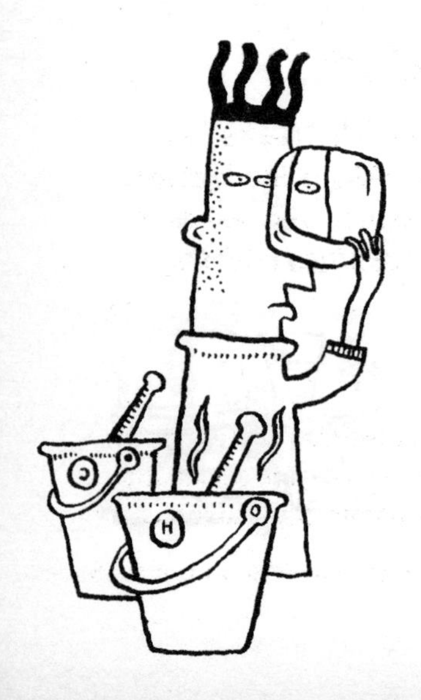
H

WHITE WATER DRINKING

Q ***Why do anisette-based drinks, such as Pernod or Sambucca, turn white when water is added to them?***

Alexander Hellemans
Amsterdam, The Netherlands

A Anisette-based drinks rely on aromatic compounds called terpenes for their flavour. These terpenes are soluble in alcohol, but not in water. The 40 per cent or so alcohol in the drink is enough to keep the terpenes dissolved, but when the drink is mixed with water they are forced out of solution to give a milky-looking suspension.

Absinthe, a similar drink which is based on wormwood and is now banned in some countries because of its toxicity, gives a more impressive green suspension. Terpenes are responsible for a lot of the harsher plant scents and flavours including lemon grass and thyme.

Thomas Lumley
Newtown, Australia

CUBED ROUTE

Q ***The pattern of bubbles in ice cubes has always fascinated me. Why is the highest concentration in the centre, with spines of progressively smaller bubbles radiating outwards?***

Carol-Ann Craig
Edinburgh

A Your correspondent has observed two effects at work in the solidification of ice—the segregation of a solution into two separate physical forms or phases on solidification, and the way in which crystals tend to grow.

Tap water contains, on average, 0.003 per cent dissolved air by weight but the solubility of air in ice is very small. When you put your ice tray in the freezer, heat flows away from the water through each of the outer faces of the cube and ice crystals containing almost no air will start to form from each of these outer surfaces. These will tend to grow inwards in the form of columnar crystals.

The unfrozen water, lying between these crystals and in the centre of the cube, gradually becomes completely saturated with the air that is driven from the earlier forming ice.

When the concentration of air in the remaining water reaches 0.0038 per cent it forms what is known as a eutectic composition—a mixture that has the lowest freezing point of all possible mixtures of that particular substance. When this occurs it all freezes at once, separating into a mixture of approximately 2.92 per cent air by volume, and ice. The smaller bubbles radiating outwards are from the eutectic which was trapped between the columnar crystals and the high concentration in the middle is where the remaining bulk of the eutectic had been trapped.

An extreme example of the formation of columnar crystals can be seen if you bite into an iced lollipop. The segregation of dissolved elements and the formation of columnar crystals also occurs in the casting of metals.

Jane Blunt
Cambridge

HOT STUFF

Q ***Is it true that hot water placed in a freezer freezes faster than cold water? And if so why does this happen?***

Ian Popay
Hamilton, New Zealand

This question was raised many years ago in *New Scientist* and never answered satisfactorily. This time we are closer to settling the controversy with answers from several people who have tried the right experiments. Counterintuitive though it may be, it does appear that hot water can freeze more quickly in a refrigerator. Better thermal contact if the water container is placed into an iced-up freezer compartment and a different pattern of convection currents which allow hot water to freeze faster seem the best explanations. Which effect predominates depends on the fridge, the container, and where it is placed—Ed.

A The questioner is correct—it is possible to produce ice cubes more quickly by using initially hot water instead of cold. The effect can be achieved when the container holding the water is placed on a surface of frost or ice. The higher temperature slightly melts the icy surface on which the container rests, greatly improving the thermal contact between the container and the cold surface. The increased rate of heat transfer from the container and contents more than offsets the greater amount of heat that has to be removed. The effect cannot be obtained if the container is suspended or rests on a dry surface.

This effect was first noted by Sir Francis Bacon using wooden pails on ice. My own investigation showed ice cubes could be obtained within

15 minutes rather than 20 minutes if the frost in the refrigerator was deep enough. The incentive to get your ice a little quicker is obviously greater in Australia than in cooler countries.
Michael Davies
University of Tasmania

But Sir Francis Bacon was not the first to note the effect. Aristotle's account in *Meteorology* below implies a similar explanation:

'Many people, when they want to cool water quickly begin by putting it in the sun. So the inhabitants when they encamp on the ice to fish (they cut a hole in the ice and then fish) pour warm water round their rods that it may freeze the quicker; for they use ice like lead to fix the rods.'
David Edge
Hatton, Derbyshire

And it seems untrue that the 'effect cannot be obtained if the container is suspended or rests on a dry surface . . .'

This question was raised in *New Scientist* in 1969, by a Tanzanian student named Erasto Mpemba. He discovered that ice cream mixture froze more quickly when put in the freezer hot than if allowed to cool to room temperature first. I got the same sceptical comments from my teachers as Mpemba did when I based my sixth form project on his question.

First, the project showed that water, either from the tap or distilled, behaved in the same way as ice cream mixture; the chemical composition is not important. Second, it demonstrated that a reduction in volume by evaporation from hot water was not the cause. Placing thermocouples into the water showed that water at about 10 °C, reached freezing point more quickly than water at about 30 °c as predicted by Newton's law of cooling, but that thereafter water that started off warm solidified more quickly. In fact, the maximum time taken for water to solidify in the freezer occurred with an initial temperature of about 5 °c, and the shortest time at about 35 °c. This paradoxical behaviour can be explained by a vertical temperature gradient in the water. The rate of heat loss from the upper surface is proportional to the temperature. If the surface can be kept at a higher temperature than the bulk of the liquid, then the rate of heat loss will be greater than from water with the same average temperature, uniformly distributed. If the water is in a tall metal can rather than in a flat dish, the paradoxical effect disappears. We argued that temperature gradients in the tall can were short-circuited by heat conduction through its metal walls.

The question has certainly made me reluctant to take accepted wisdom for granted when it comes to observations which do not fit preconceived notions of what is correct.

J. Neil Cape
Penicuik, Midlothian

The classic experiment uses two metal buckets placed in the open air on a cold, preferably windy, night. Stationary water is a poor conductor of heat and ice forms on the top and around the sides. If the initial temperature is around 10 °c, cooling of the core is very slow, particularly as loose ice floats to the top, inhibiting normal convection. There is no means by which the warmer water can come into contact with the cold bucket and transfer its energy to the outside.

If the initial temperature is about 40 °c, strong convection is established before any water freezes, and the entire mass cools rapidly and homogeneously. Even though the first ice forms later, complete solidification of the hot water can occur more quickly than if the water starts off cold.

The conditions are critical. Obviously, if the cold bucket starts at 0.1 °c and the hot at 99.9 °c the experiment is unlikely to cause surprise. The containers must be large enough to sustain convection with a small temperature gradient, but small enough to extract heat quickly from the bucket's surfaces. Forced air cooling on a windy night helps.

It is difficult to generate suitable conditions in a domestic freezer but the anomaly can be demonstrated in an industrial chiller or a laboratory environmental chamber.

Alan Calverd
Bishop's Stortford, Hertfordshire

It's true and I have verified the assertion by experiment. The only limitation is that the container of water must be relatively small so that the capacity of the freezer to conduct away the heat content is not a limiting factor.

Cold water forms its first ice as floating skin, which impedes further convective heat transfer to the surface. Hot water forms ice over the sides and bottom of the container, and the surface remains liquid and relatively hot, allowing radiant heat loss to continue at a higher rate. The large temperature difference drives a vigorous convective circulation which continues to pump heat to the surface, even after most of the water has become frozen.

Tom Hering
Kegworth, Leicestershire

This is a cultural myth. Hot water will not freeze faster than cold water in the freezer. However, hot water cooled to room temperature will freeze faster than water that has never been heated. This is because heating causes the water to release dissolved gases (mostly nitrogen and oxygen) which otherwise reduce the rate of ice crystal growth.

Tom Trull
University of Tasmania

Sceptical Tom Trull from the University of Tasmania might like to stroll over to the refrigerator of our first letter writer, Michael Davis, also from the University of Tasmania. The experimental evidence suggests that the effect is real—the absence of dissolved gas could be another factor that speeds crystal growth.

And there could be yet another factor that none of our letter writers has described—supercooling. New research reported in *New Scientist* (4 May 1996) shows that because water may freeze at a variety of temperatures, hot water may begin freezing before cold. But whether it will completely freeze first may be a different matter—Ed.

In scientifically controlled experiments this effect seems to be real. We assume that the temperature in the freezer stays constant during the freezing process as do the variables of the samples such as container size, conduction, and convection properties inside and outside the container.

However, I feel that one more variable is present and that is an overlooked temperature variation in the freezer. The temperature oscillation inside the freezer depends on the sensitivity of the thermoelement and the timer of the controller system. We may assume that at the freezer's standard temperature the power used for cooling the freezer operates at a standard rate. If a bucket of cold water is added, it may produce only a small effect on this power output because it will not trigger the temperature sensor. However, a bucket of hot water may easily activate the sensor and release a short but powerful cooling of the freezer with a cooling overshoot depending on the timer.

This may be overlooked by an observer at home. I have seen a similar effect in an electric sauna. By fooling the temperature sensor by splashing water I increased the oven's output.

Matti Jarvilehto
University of Oulu, Finland

TRANSPARENT ROCKS

Q ***How do you get transparent ice? Ice from my freezer always contains bubbles. I've used filtered and boiled water but the ice is never like that seen in ads for Scotch.***

Philip Susman
Monash University, Victoria

A Ice made in domestic freezers is inevitably cloudy because of the dissolved air that tap water contains (around 0.003 per cent by weight). As the water in the ice trays drops below freezing point, crystals form around the edges of the compartments. These are pure ice and they contain very little air because the solubility of air in ice is very low and the liquid left behind can still hold it in solution.

Once the concentration of air in the liquid reaches 0.0038 per cent by weight and the temperature has dropped to −0.0024 °c, the liquid can contain no more air and a new reaction begins. As the water freezes the air is forced out of solution. The natural state of air at the temperature and pressure involved here is a gas, so it forms bubbles in the ice.

Commercial ice machines produce attractive, clear ice by passing a constant stream of water past freezing metal fingers, or over freezing metal trays. This freezes out a fraction of the water while the rest of it is discarded before the concentration of air gets too high. When the ice is thick enough the metal fingers or tray are warmed to release their crystal clear ice that is good enough to film.

Alas, without an ice machine, the questioner may have to make do with cloudy ice cubes.

Andrew Smith
Newcastle-upon-Tyne

Water has its highest density at around 4 °C. Below that the water gets less dense as it approaches freezing point.

Air bubbles form in the ice when the cooling of the water is too rapid, which causes one part of the water to be at a different temperature to other parts. Ice is usually formed at the top of the water first because the warmer and denser water sinks beneath the ice layer that begins to form there.

Additionally, the top layer is usually the part that is in contact with the cold environment. This is similar to what happens in a frozen lake. The various expansion rates of different parts of the water will inevitably create air bubbles that cannot escape because of the ice sheet above.

To avoid ice bubbles, the trick is to cool the water very slowly so that there is no large temperature gradient which can cause widely different expansion. Cooling it slowly also allows air to have sufficient time to move through the liquid and evaporate before it is trapped by the solid ice.

Han Ying Loke
Edinburgh

Your failure to achieve clear ice even after boiling the water may depend on what precautions you took in addition to filtration and boiling. For instance, gas can still be dissolved if water is poured after boiling has finished and hard water may also need deionizing to rid it of gas.

It also helps to exclude air while the water is cooling, perhaps with clingfilm. Try to achieve zone freezing by cooling the water slowly from the top down, maybe in a polystyrene container with only clingfilm over the top. Do not use a vacuum flask because the glass is far too fragile.

Although the amount of gas that freezes out is unaffected by the method of freezing, zone freezing starts by giving a nice, thick chunk of clear ice. As freezing progresses murky ice appears and then you can stop the process.

Jon Richfield
Dennesig, South Africa

A Water contains dissolved gases. When it freezes, the gas is forced out of it to form bubbles that are then trapped in the ice, making it look opaque.

In order to get transparent ice you should use warm rather than cold water, because warm water contains less dissolved gas. Also, try to reduce the power of your freezer in order to allow time for the gases to diffuse out just before freezing. I have tried this and it works very well.

Gabriel Souza
Cambridge

A I'm afraid that your correspondent has been seduced by a professional photographer who uses hand-carved perspex 'ice' cubes in adverts for Scotch because they don't melt under the studio lights. If he looks carefully, he might also see the tiny glass 'bubbles' they use on the meniscus of other

drinks—these won't disappear at the wrong moment either.
Martin Haswell
Bristol

WAYWARD WATER

Q ***Why does an electrostatically charged plastic comb attract a stream of water? Try the experiment in your bathroom. Turn on the tap until you get a thin, steady flow of water. Comb your hair and place the comb near the point where the water emerges from the tap. Large deflections can occur. What property of water is involved in this phenomenon?***
David Doff
Trinity College, Dublin

A The molecules of water are electrically neutral but, like the molecules of other dielectric substances, they are characterized by regions of positive and negative charge. The electrical field caused by the comb (produced by an excess of electrons collected by the comb from the friction caused by combing) attracts the more positive parts of each water molecule and repels the negatively charged regions.

Because water is not a metal, the charges are not free to migrate through the dielectric substance in the direction of the field. Instead, the molecules are rotated and stretched by the field, and tend to become aligned head-to-tail along the electrical field-lines, with their positive ends pointing towards the comb.

The charges on each molecule therefore neutralize each other, except at the water-air interface that is closest to the comb, and also at the interface farthest from it. In these regions, they are not paired with opposite charges, so they build up areas of equal positive and negative charge on the opposing sides of the cylindrical water surface. The stream is now polarized, and it is this property of polarization that allows dielectric materials to be attracted to a source of electrical field.

If the field caused by the comb is weaker where the distance is greater (that is, the far side of the water stream) it will repel the more remote negative charge less strongly than it attracts the positive charge, and there will be a resultant force on the stream of water that deflects it towards the comb. In a completely uniform electrical field caused, for example, by this experiment being performed using an infinitely large comb in a huge bathtub, there would be no net attraction of the water stream. This situation is difficult to achieve experimentally because the force on the stream depends on the gradient of the square of the field, which is sensitive to a lack of uniformity in a very small field.

Water is by no means unique in showing what is called the Doff effect. Try it with black treacle, in the kitchen sink. In a thin filament of treacle dribbled from a tablespoon, the flow is slowed by the high viscosity of the fluid, so that the electrical field has more time to act on each section of the flow's length. The resulting deflection is spectacular.

Roger Kersey
Nutley, East Sussex

The comb will have another effect. When the stream of water comes out of the tap, it will eventually break up into drops. A charge will change the droplet size, with a weak charge causing bigger drops than normal. This is caused by the weak charge on the drops on opposite

sides of the stream attracting each other. Where a strong charge is involved, the repulsion in a single drop tears it apart.
Edward Edmondson
Gloucester

FLOATING ON AIR

Q ***I have been told that if I lie on an air mattress near to the shore, it will always tend to drift out to sea, even when there is no wind. Is this true? If so, why?***
Angus Cameron
Canada

A It is not necessarily true. It depends on the combination of currents, wave patterns, and, most markedly, the wind. However, some things do bias towards a seaward drift. Incoming surf has little impact on a smooth, light, high-floating bubble like an air mattress, so it cannot give the mattress much of a push. Increasing the wave's speed does not improve the 'grip' on the mattress. And the faster the mattress is moved, the greater the air resistance on it.

Long curling waves (combers) tend to come in fast and high and, to a lesser extent, so do broken waves as they roll in. Such waves pass quickly, so they have only a brief effect. If you watch an unoccupied air mattress, you will see that most waves pass beneath without giving much beachward acceleration.

The backwash is slower and lower than the incoming wave, and lasts longer, so it drags the mattress seaward rather more effectively. Beyond the backwash area the effect dies away and the mattress drifts more or less with the wind and currents.

If the mattress is ridden by someone denser than air, the seaward drift effect is decreased, because the mattress floats deeper and the waves get a better grip. In fact, a skilled mattress-rider can surf inwards by adjusting the mass distribution so as to take advantage of the force of the incoming wave.

But, in general, the mattress will always go the wrong way, especially if you are not a strong swimmer nor a skilled surfer.

Even onshore winds are not to be relied on—in some areas they can switch directions within minutes.

Jon Richfield
Dennesig, South Africa

A This is an example of a constrained random walk. Imagine that the mattress is moved completely at random by the wind, waves, and current. Over time, the result of these small random movements is a net displacement in a random direction. (Probability theory shows that the distance travelled increases with the square root of the time taken.) Because the mattress starts next to the shore, it can only move so far towards the shore. In effect, it 'bounces' off the shore. So the net result is a drift out to sea, even if there is no seaward wind or current.

Andy Kilpatrick
Enfield, London

BLOWING BUBBLES

Q ***Why is it that when you have a bubble bath, all the bubbles disappear when you***

use ordinary soap? Is it possible to stop this from happening?

MARI WILLIAMS
LONDON

Both bubble bath and soap are surface-active agents (surfactants) but they are normally of opposite and incompatible types, which is why the bubble bath collapses. Surfactant molecules tend to be shaped like tadpoles with elongated tails. The head is water-attracting and may carry an electrical charge which can be either positive (cationic) or negative (anionic), while the tail is grease-attracting, water-repelling, and carries no charge. The foaming and washing properties of these surfactants result from this characteristic structure.

The bath problem arises because cationics are usually chosen for bubble-bath mixtures because of their high foaming characteristics, whereas soaps which have good washing characteristics are anionics.

When the two come together, the oppositely electrically charged heads are strongly attracted to each other, and the resulting entity is a 1:1 complex of the two surfactant molecules. Such structures have neither good foaming nor good washing characteristics, and the foam collapses. Fortunately for the bather, the soap tends to be used in excess and so its effect is not totally lost.

This different surfactant effect is also present in hair conditioners and shampoos, which is why two-in-one formulations have only recently become available. Any direct attempt to mix them would result in a sludge devoid of shampooing or conditioning properties.

In the new two-in-one shampoo and conditioner formulations a number of patented tricks are used, essentially to keep the conditioner separate from

the shampoo until the dilution effect of rinsing out the shampoo releases it.
John Brennan
Wassenaar, The Netherlands

Bubbly Attraction

Q ***I have noticed that an individual soap bubble generated by the action of my dripping bathroom tap is content to spend a solitary (and brief) existence drifting haphazardly on the surface of the water. However, when approached by other like-minded bubbles a sense of urgency seems to draw them together. It happens when they approach within a distance of a bubble's diameter, when they suddenly accelerate towards each other and join together in a double or multi-bubble mass.***

This also happens when bubbles approach the side of the bath (or for that matter, any human parts which happen to be projecting out of the water). What is at work here?
Ian Houston
Longfield, Kent

A The agent responsible for these effects is surface tension, although the process of bubble aggregation may be separated into two distinct phases. First, a solo bubble has an internal air pressure slightly greater than that outside. This is caused by the squeezing effect of surface tension, which acts to reduce the bubble's surface area. As a result, the water level inside the bubble behaves like a tiny float, rising to the highest available level. In the vicinity of the bubble the outside water surface is pulled into a rising curve, the meniscus, again by surface tension. A bubble encountering the gradient will float up the slope.

Once the two bubbles come into contact, surface tension will force the composite bubble into the configuration with the smallest surface area which, for bubbles of similar size, resembles two quarter spheres joined along a straight partition. If bubbles coalesce in the centre of a calm body of water the energy released by their fusion may be seen propagating outwards as a faint circular ripple.

Wherever you find a meniscus you will find a source of attraction for bubbles, so the sides of the bath and other objects will be a focus for their attentions. Because the meniscus does not extend far into the water the force between bubbles has a short range.

Tommy Moorhouse
Durham

Consider a lone bubble on a large body of water. The distribution of surface tension around it is uniform (hence it is round) with the tension level reducing the closer you are to the bubble. Because the surface tension is uniform around the bubble, it does not impart any forces on the bubble which drifts according to any momentum given to it on formation and by fluctuations in air pressure.

Should two bubbles pass close to each other such that their tension distributions overlap, the equilibrium is destroyed. The result is an area of low tension between the bubbles which the high tension elsewhere tries to eliminate. This is done by pushing the bubbles together whereupon a new equilibrium is reached.

R. A. Best
Birmingham

OVER THE TOP

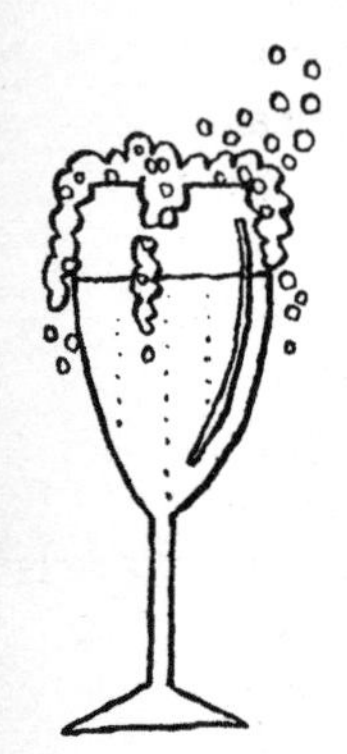

Sparkling wine or beer poured into a dry glass froths up. If the glass is wet this does not happen. Pour some sparkling wine into a glass so that it froths up to the rim, let the bubbles subside and you can then pour the rest of the wine quickly, knowing it will not froth over the top. Why?

H. Sydney Curtis
Hawthorne, Queensland

Beer, sparkling wine, and other fizzy drinks are liquids which are supersaturated with gas. Although thermodynamics favours the gas bubbling out of the dissolved state, bubble formation is unlikely because bubbles must start small.

The pressure of these tiny bubbles can reach about 30 atmospheres in a bubble only 0.1 micrometres in diameter. Because the solubility of gases increases with increasing pressure (Henry's law) the gas is forced back into solution as quickly as it comes out.

Bubbles can form around dust particles, surface irregularities, and scratches. These nucleation sites are hydrophobic and allow gas pockets to grow without first forming tiny bubbles. Once the gas pocket reaches a critical size, it bulges out and rounds up into a properly convex bubble whose radius of curvature is sufficiently large to prevent self-collapse . . .

D. P. Maitland
Department of Pure and Applied Biology, University of Leeds

. . . In addition, there is a cascade effect. If the quantity of bubbles reaches a certain critical number per unit volume, this in itself constitutes a physical disturbance and results in the release of yet more bubbles.

Nucleation may be precipitated by a variety of imperfections. Minute crystals of salts (such as calcium sulphate) may remain if the glass has been left to dry by evaporation after being washed in hard water. Tiny cotton fibres may be left behind if the glass has been dried with a tea cloth. Dust particles may have settled on the glass if it has been left standing upright for any length of time. And tiny scratches will be present on the inside surface of all but brand-new glasses.

Once the inside of the glass is wet, any salt crystals will have dissolved and any cotton fibres will no longer function as centres of nucleation. Most of the dust particles and all of the scratches will, of course, still be there. However, these will have been coated with liquid and the fresh carbonated liquid can only reach them very slowly, by diffusion. Bubbles will still be produced, but at a rate that is too slow for the cascade effect to come into play. As a result, the drink will not froth over.

Allan Deeds
Daventry, Northamptonshire

To demonstrate the above, take a glass, and thoroughly coat the inside with an oil, which is a more efficient surface covering agent than water. Then add a less expensive carbonated drink such as lemonade. The effervescence will be nil or minimal. Add a few million centres of nucleation from a large spoonful of granulated sugar and the effervescence will be volcanic.

Ronald Blenkinsop
Westcliff-on-Sea, Essex

Thanks to modern production techniques, today's glasses are of such good quality that some manufacturers build in deliberate imperfections, especially in beer glasses, in order to generate enough bubbles to maintain the head on the top of your tipple.

Tony Flury
Ipswich, Suffolk

YOUR BODY

CATCH YOUR DEATH

Q ***Is there any connection between being cold and catching a cold? If not, why is there so much folklore about catching a cold if you sleep uncovered or in a draught?***
Antonis Papanestis
London

A No, there is no connection. The erroneous association developed for several reasons.

The viruses that cause colds spread faster in the winter because people spend more time inside, where they are closer together.

People close the windows in winter so air contaminated by virus particles is not diluted by 'fresh' air from outdoors. This makes it easier for the virus to spread.

The cold, dry air of winter makes the mucous membranes in the nose swell. This produces the 'runny nose' we often incorrectly associate with an infection caused by a cold virus.

The experience of catching a chill and getting a cold is actually the reverse of the correct order of things. The 'chill' is often the first sign of fever that is the result, not the cause of, the infection by the cold virus.
Mark Feldman
Northland, New Zealand

A Studies have shown that there is no correlation between environmental temperature and suffering from colds. The origin of the folklore that predicts colds, flu, or pneumonia after being exposed to cold is the short period of fever that precedes the distinctive symptoms of these illnesses. These periods of fever make the patient feel cold and shivery. Shortly after developing other symptoms, the patient then associates the illness with having 'caught cold'. Indeed, the flu is called influenza from the belief that it was caused by the 'influence' of the elements. The fact that isolated researchers living in Antarctica never catch colds confirms that these are caught from people and not from 'cold'.

Pedro Gonzalez-Fernandez
London

A There is actually less chance of you catching a cold in the cold. The virus known as the common cold dies in cold and needs warmth (say the cosy indoors of a home beside the fire that was started to keep out the cold) to thrive.

Esperandi
No address supplied

LIGHT SNEEZE

Q ***I have noticed that many people tend to sneeze when they go from dark conditions into very bright light. What is the reason for this?***

D. Boothroyd
Harpenden, Hertfordshire

A Photons get up your nose!

Steve Joseph
Sussex

A I think that the answer may be fairly simple: when the Sun hits a given area, particularly one shielded or enclosed in glass, there is a marked rise in local temperature. This results in warming of the air and a subsequent upward movement of the air, and, with it, many millions of particles of dust and hair fibres. These particles quite literally get up one's nose within seconds of being elevated; hence, the sneezing.

Alan Beswick
Birkenhead, Merseyside

My mother, one of my sisters, and I all experience this. I feel the behaviour is genetic and confers an unrecognized evolutionary advantage. I have questioned many people, and we Sun-sneezers seem to be in the minority. However, as the ozone thins and more ultraviolet light penetrates the Earth's atmosphere it will become increasingly dangerous to allow direct sunlight into the eye. Those of us with the Sun-sneeze gene will not be exposed to this, as our eyes automatically close as we sneeze! The rest of the population will gradually go blind, something not usually favoured by natural selection.

Alex Hallatt
Newbury, Berkshire

The tendency to sneeze on exposure to bright light is termed the 'photic sneeze'. It is a genetic character transmitted from one generation to the next and which affects between 18 and 35 per cent of the population. The sneeze occurs because the protective reflexes of the eyes (in this case on encountering bright light) and nose are closely linked. Likewise, when we sneeze our eyes close and also water. The photic sneeze is well known as a hazard to pilots of combat planes, especially when they turn towards the Sun or are exposed to flares from anti-aircraft fire at night.

R. Eccles
Common Cold and Nasal Research Centre, Cardiff

Here are some early thoughts on the subject of light sneezing from Francis Bacon's *Sylva*

Sylvarum (London: John Haviland for William Lee, 1635: 170):

Looking against the Sunne, doth induce Sneezing. The Cause is, not the Heating of the Nosthrils; For then the Holding up of the Nostrills against the Sunne, though one Winke, would doe it; But the Drawing downe of the Moisture of the Braine. For it will make the Eyes run with Water; And the Drawing of Moisture to the Eyes, doth draw it to the Nosthrills, by Motion of Consent; And so followeth Sneezing; As contrariwise, the Tickling of the Nosthrills within, doth draw the Moisture to the Nosthrills, and to the Eyes by Consent; For they also will Water. But yet, it hath been observed, that if one be about to Sneeze, the Rubbing of the Eyes, till they run with Water, will prevent it. Whereof the Cause is, for that the Humour, which was descending to the Nosthrills, is diverted to the Eyes. [*sic*]

C. W. Hart
Smithsonian Institution, Washington DC

COMES IN HANDY

Q ***Why do we have fingerprints? What beneficial purpose could they have evolved to serve?***
Mary Newsham
London

A Fingerprints help us in gripping and handling objects in a variety of conditions. They work on the same principle as the tyres of a car. While smooth surfaces are fine for gripping in a dry environment, they are useless in a wet one. So we have evolved a system of troughs and ridges, to help channel the water away from the fingertips, leaving a dry surface which allows a better grip. The unique pattern is merely a useful

phenomenon that is used by the police to identify individuals.
James Curtis
Bradford, West Yorkshire

A Fingerprints are the visible parts of rete ridges, where the epidermis of the skin dips down into the dermis, forming an interlocking structure (similar to interlaced fingers). These protect against shearing (sideways) stress, which would otherwise separate the two layers of skin and allow fluid to accumulate in the space (a blister). They appear on skin surfaces which are subject to constant shearing stress, such as fingers, palms, toes, and heels. The unique patterns are simply due to the semi-random way in which the ridges and the structures in the dermis grow.
Keith Lawrence
Staines, Middlesex

Crinkle tips

Q ***Why does skin—especially of the fingers and toes—become wrinkled after prolonged immersion in water?***
Lloyd Unverfirth
Wahroonga, Australia

A The tips of fingers and toes are covered by a tough, thick layer of skin which, when soaked for a prolonged period, absorbs water and expands. However, there is no room for this expansion on fingers and toes, so the skin buckles.
Steven Frith
Rushden, Northamptonshire

A Your whole body does not become crinkled as the skin has a layer of waterproof keratin on the surface, preventing both water loss and uptake. On the hands and feet, especially at the toes and fingers, this layer of keratin is continually worn away by friction. Water can then penetrate these cells by osmosis and cause them to become turgid.

Robert Harrison
Leeds, West Yorkshire

TAKE THE PILS

Q ***Why is it that when I walk home from the pub after a few beers, I always stumble to the left more than to the right?***

Chris Wood
Liverpool

A A similar situation arises when people wander in the forest or desert. Although they may intend to walk in a straight line, if they are lost and have no landmarks to guide them, most people will unconsciously walk slightly towards the left, making a big anticlockwise circle bringing them back to their starting-point.

The reason for this is that most people have a slightly stronger and more flexible right leg. This is common knowledge among sports scientists, and most people who have undergone strength tests in their legs can confirm it.

Most people also find they can lift their right leg slightly higher than their left. The right leg has a longer stride than the left one and so when there are no guiding landmarks a circular walking route is the result.

Also, the slightly greater strength of the right leg means that when you push on the ground with your right foot, the push to the left is slightly greater than the push to the right produced by the left foot. The longer stride and greater push combine to cause most people to move in an anticlockwise manner in the course of a long walk.

Han Ying Loke
Edinburgh

The human body is never perfectly symmetrical. In this case, the right leg seems to be longer than the left. A beer mat placed in the left shoe underneath the foot should remedy the problem quite easily.

J. Jamieson
Marlow, Buckinghamshire

Everyone has a dominant eye which they rely on more than the other, weaker eye. Instinctively, we try to walk where we can see best (although we normally correct this to allow us to walk forwards). So, when we stumble, it is more likely that we will stumble in the direction of our dominant eye.

This is because the brain, in trying to recover the situation, has to react fast and gives more weight to the information coming from the dominant eye to work out where to put the feet in order to regain balance. Hence the feet tend to be aimed at a position towards the side of your body on which the dominant eye lies, resulting in a stumble in that direction. In this case the questioner's dominant eye is obviously his left.

This phenomenon can be used to steer riding animals—simply cover up one of their eyes and

they will tend to move in the direction of their remaining eye.
Adrian Baugh
Shrewsbury, Shropshire

The questioner obviously walks to the pub with his change in his right pocket and his keys in his left. After spending all his money on beer the weight of his keys pulls him to the left as he walks home.
Simon Thorn
Perth, Tayside

Members of the Department of Physics at Auckland University have held consultations regarding this issue and our most popular theory derives from an application of the simple principles of gravity gathered from our common experience in returning from pubs in Auckland.

Currency in denominations lower than NZ$10 is mostly in coins, some of them quite large in size. During an evening in the pub, the drinker accumulates a large number of such coins in his or her pocket. Assuming that English coinage is similar and that your questioner habitually carries his coins in his left pocket, elementary laws of gravity dictate that his gait will incline to the left. It is not uncommon for some New Zealanders in similar circumstances actually to walk in a circle.
Nelson Christenson
University of Auckland

After standing for endless hours in a pub with your beer glass in your right hand, it is inevitable that you are still subconsciously counterbalancing the glass's weight, and thus stumbling more to the left. The opposite can be demonstrated in left-handed beer drinkers.
No name or address supplied

IT AIN'T HALF HOT

Q ***Why does our temperature go up when we are ill?***

KERON BAGON (AGED 10)
RADLETT, HERTFORDSHIRE

A The question of why our temperature goes up during illness can be split into two parts. First, you need to know what makes the temperature go up, and secondly, what advantage an increase in temperature offers.

The increase in core temperature observed during illness is commonly called fever and occurs in response to infection by a pathogenic organism or certain types of physical injury.

For example, when a person becomes infected with bacteria, the white blood cells of the immune system recognize the incoming pathogen as foreign and initiate the first stages of the immune response—the acute phase. In this reaction, white blood cells called monocytes release a variety of proteins called cytokines. These are central to the immune and inflammatory response.

In particular, there is a predominance of two types of cytokine called interleukin-1 (IL-1) and tumour necrosis factor-alpha. These cytokines are known as pyrogenic because they cause an increase in body temperature. It is not clear how they induce fever, but it is known that they also cause the production of other chemicals in the brain. The main group of chemicals produced in this effect are the prostaglandins. These react very strongly with the hypothalamus area of the brain, which then sends a signal to the body to increase the temperature. The mechanisms that the brain employs to effect this are not certain but are known to include increasing the metabolic rate

and inducing shivering. These two processes burn metabolic fuel faster than normal, and body heat is given off.

The question of what advantage fever confers is interesting. Experimental work shows that the mortality of animals decreases if the fever is untreated, and that elevated temperatures can enhance certain aspects of the immune response. Furthermore, the growth rates of various types of bacteria are slowed at temperatures above normal body temperature. Indeed, the ancient Greeks believed that fever was beneficial; even in this century fever has been used to treat certain illnesses. For example, syphilis used to be notoriously difficult to treat, so doctors gave their patients malaria which fought the syphilis, knowing that they could get rid of the malaria later.

Nigel Eastmond
University of Liverpool

With a few exceptions, our body temperature rises in response to infection. Our immune system becomes activated and seeks to destroy the source of infection. The macrophages produce a protein signal called IL-1, which triggers the destruction of the infection and is also responsible for elevating the body's temperature.

This has three possible functions: raised body temperatures may affect the ability of micro-organisms to replicate; T cells work best at temperatures between 38 °c and 40 °c; during fever, the level of iron circulating in the body falls and micro-organisms need iron in order to replicate.

So it is to our advantage that our body temperature rises when we have an infection. Incidentally, many of the symptoms we experience when we have an infection—tiredness, fever, sleepiness, aching joints, and a lack of appetite—are due to IL-1 and its related proteins, and not the micro-organism. It is our body's way of making us slow down so we can recover from an infection and in most cases it works.

Janet Reid
St James's Hospital, Leeds

REPEAT THAT

Q **Do hiccups serve any useful purpose? And do other creatures have them?**

N. de Bray
Bristol

A A hiccup or singultus is a neurological reflex, which involves the phrenic and vagus nerves and the medulla of the brain. It has no known physiological function. However, it is a normal and frequent event in fetuses and may continue to afflict newborn babies for a short while after birth. So perhaps it has something to do with the fetus in effect living underwater while in the womb, and is merely an annoying vestige in adults.

Low levels of carbon dioxide in the bloodstream make hiccups worse, which suggests that their function may be to control respiration.

Holding the breath presumably stops hiccups by raising the concentration of CO_2 in the arteries. Inhaling CO_2 has also been used to treat them. There are many other folk cures, including drinking iced water, swallowing three times without drawing breath (both of which, like sneezing, interrupt the reflex by stimulating the pharynx), pulling the forefingers, swallowing mint, and being subjected to a sudden fright.

Massaging the lower end of the oesophagus using an endoscope has been reported as effective in some cases. Surgical crushing of the phrenic nerve has also been used but it is not recommended. In intractable cases drugs may help. The wide variety of proposed treatments for hiccups suggests that there is no universal remedy.
Jeffrey Aronson
Radcliffe Infirmary, Oxford

A Not only humans suffer from hiccups. I observed that my cat sometimes gets them. They are heavy jerks in intervals of five seconds or less. They never last longer than a minute and vanish without him drinking a glass of water or holding his breath and counting to seven.
Joachim Henkel
Langenbach, Germany

POINTLESS

Q ***Ever since the recent birth of my son, who I breastfeed, I have wondered why men have nipples.***
Dorthe Mindorf
Aalborg, Denmark

A Many suggestions for this phenomenon have been offered. They may exist to help men check that their vests are on straight, or be present as a safety feature—to warn us how far out from the beach we can safely wade.

However, there is a more plausible explanation. Male and female human embryos are identical in the early stages of their development. If the fetus receives a Y chromosome from its father, a hormonal signal is produced: the labia fuse to

form a scrotum, the gonads develop as testicles and a male results. Otherwise the 'default' female remains.

Various structures in the adult reflect the symmetry of male and female and their common embryonic source. Men have nipples because they have already begun to develop when the 'switch to male' signal is received. The development of breasts is halted in most—but not all—cases but the nipples are not reabsorbed.

Another effect of these developmental pathways which are shared by both males and females is pointed out in Stephen Jay Gould's essay 'Male Nipples and Clitoral Ripples' (which can be found in the Penguin 60th anniversary collection, *Adam's Navel*). Males need plenty of blood vessels and nerve endings in their penises to achieve erections. Because the penis and clitoris have their origins in the same structure, females have the same number of blood vessels and nerve endings packed into a much smaller area, resulting in the enhanced sensitivity of the clitoris.

Conclusive evidence that God is not a man?

Jim Endersby
University of New South Wales

By the left

Q ***Why is it that when two people walk together they often subconsciously start to walk in a synchronized manner. Is this some natural instinct?***

Simon Apperley
Cheltenham, Gloucestershire

A The zoologist and specialist in human behaviour, Desmond Morris, says that the reason that people start to walk like each other is that they have a subconscious need to show their companion that they agree with them and so fit in with them. This is also a signal to other people that 'we are together, we are synchronized'.

Other studies suggest that we adopt the mannerisms of our companions as well, especially our superiors, such as crossing our legs in the same directions as others. An example often given is when, in a meeting, the boss scratches his nose and others at the table then follow him without realizing it.

Adithi
Hong Kong

While it is purely unsubstantiated opinion, I do have an answer to why people tend to synchronize their steps. Observing a group of children walking in a park recently, supervised by two adults, I noted that the adults synchronized their steps and direction, while the children walked, ran, and skipped apparently at random, running ahead, lagging behind, and deviating from the common course.

Perhaps these children, unpolluted by society's emphasis on conformity, have not yet learned that it is unacceptable to march to your own drum.

Todd Collins
Wagga Wagga, Australia

The next time you walk alongside somebody, walk out of step. Then try to follow the conversation you are having. You will soon fall back into step, because once you are in step with the other person, it is easier to watch where you are walking and then turn to look at them.

Communication is easier with another person when you are in close proximity and when both

faces are relatively stable and not bobbing all over the place.
Hamish
No address supplied

A Here is a more prosaic (less sociologically inclined) explanation. When people walk they have a slight side-to-side sway. Two people walking together and out of step would bump shoulders every second step.
Peter Verstappen
Kaleen, Australia

Blind blow

Q ***Why do we close our eyes when we sneeze?***
Bridget Perez
No address supplied

A To stop our eyes popping out.
Maria Taylor
Kumamoto, Japan

A The force of a sneeze is transmitted through the nasal passages which run from your nose to your eyes. If you smoke, you can try this experiment: inhale a cigarette, hold your nose and mouth shut, and push hard. The smoke will come out through your tear ducts.
James Merwin
No address supplied

A It's not just flimsy eyelids that hold in your eyes during the force of a sneeze. The muscles around your eyeballs also contract and effectively form a solid barrier.
Shaun Lawton
No address supplied

A If you were to sneeze with your eyes open, you would blow your eyeballs out. This has been

known to happen among people who prop their eyes open with toothpicks to stay awake.
Monica Baird
No address supplied

This last story is surely a modern myth. If anyone knows of someone who blew their eyes out, we'd like to hear. Don't try blowing smoke out of your eyes, and don't try the same trick after drinking milk to produce creamy tears—it works but could be dangerous—Ed.

SCREEEEECH!

Q ***Why do sounds like scratching a black-board and scraping metal make some people cringe or shiver? And how does the mind remember sounds? If it can remember sounds accurately why do we not cringe when we remember these sounds?***
Ben Cronin
London

A The danger warning sounds emitted by some of the great apes are of a similar frequency and tone to the sound made by fingernails dragged on a blackboard, so possibly a primal instinct for danger is being stimulated.
Howard Vogel
No address supplied

A People do cringe at the memory of sound. Upon reading this question, my back seized up at the thought of fingernails on a blackboard.
Tiffany Caron
No address supplied

HELPLESS LAUGHTER

Q ***Why is it that if you tickle yourself it doesn't tickle, but if someone else tickles you, you cannot stand it?***

DANIEL (AGED 7) AND NICOLAS (AGED 9) TAKKEN
WAGENINGEN, THE NETHERLANDS

A If someone was tickling you and you managed to remain relaxed, it would not affect you at all. Of course, it would be difficult to stay relaxed, because tickling causes tension for most of us, such as feelings of unease due to physical contact, the lack of control, and the fear of whether it will tickle or hurt. However, some people are not ticklish—those who for some reason do not get tense.

When you try to tickle yourself you are in complete control of the situation. There is no need to get tense and therefore, no reaction. You will notice the same effect if you close your eyes, breathe calmly, and manage to relax the next time someone tickles you.

The laughter is the result of the mild state of panic you are in. This may be inconsistent with 'survival of the fittest' theories, because panic makes you more vulnerable. But, as in many cases, nature is not necessarily logical.

SIGURD HERMANSSON
STOCKHOLM, SWEDEN

THE BEARDIES

Q ***Why does hair grow continuously on a man's face and on the heads of both sexes, but only***

to a fixed, yet renewable, length on other parts of the human body?

David Scott
London

All hair strands have a predetermined length (wear and tear excluded). This arises from the fact that hair grows at a fixed rate for a fixed duration of time. There are great variations in this process. My hair has reached 75 centimetres, yet other people have much shorter hair. After this growth period, hairs fall out and the cycle starts again. This results in an equilibrium, providing hair with a maximum length. For other types of hair, these factors are different, resulting in other maximum lengths.

Eirik van der Meer
Porsgrunn, Norway

Hair growth is cyclical, showing three distinct stages: an active period of hair fibre production called anagen followed by a period of regression called catagen, leading to a dormant stage known as telogen. After this a hair follicle returns to anagen, sheds any old fibre and produces a new hair.

Although hair follicles on our scalps seem to grow continuously they actually cycle very slowly, with anagen lasting on average between 6 and 10 years and telogen between 30 and 90 days. Rate of hair growth varies with age, sex, and the site of the hair follicle, but has a minor part to play in limiting maximum hair length. It is the duration of anagen that is most important—a brief time period results in shorter, but renewable hair fibres. What dictates the speed of cycling is not known for sure but there are several hypotheses.

Hair follicle growth can be affected by the production of hormone-like chemicals called cytokines. Numerous cytokines are produced throughout our bodies by different cells and within hair follicles themselves. Some promote hair growth and others prohibit it—leading to the suggestion of an on/off switch for hair follicles in which high levels of stimulatory cytokines initiate anagen but also promote the production of inhibitory cytokines.

When these inhibiting cytokines reach a certain concentration, the hair follicle is switched off and does not start again until the promoter cytokines accumulate at high enough levels to induce anagen. Larger scalp hair follicles may need greater concentrations of cytokines to trigger the switch and so cycle more slowly.

In experiments, the duration of anagen can be reduced or increased. Injections of cytokines called EGF and/or TGF can induce the end of anagen and the onset of catagen. EGF injections to depilate sheep as an alternative to shearing has been researched extensively in Australia, but the work remains experimental as bald sheep are prone to sunburn...

Anagen can also be prolonged using an immunosuppressive drug, cyclosporin. Normally given to organ transplant patients to prevent rejection, cyclosporin users often develop hair growth as a side effect, and larger hair follicles with prolonged anagen duration. Cyclosporin inhibits production of certain cytokines and so may disrupt chemical communication within hair follicles.

Macrophage immune cells, which produce many cytokines, congregate around hair follicles in large numbers when they leave anagen and enter catagen and could affect the rest of the immune system, which may have an influence on hair follicle cycling.

Although hair follicles look the same, there is some evidence that in embryological development different regions of hair are derived from different cell populations—a phenomenon known as 'skin mosaicism'. Our scalp hair is made up of at least two distinct populations, one within an area around our ears and the other on top of our heads.

This has already tempted some to suggest these cause variable rates of cycling in different regions of the body and different susceptibility to hair loss. Further, different hair follicles express variable concentrations of hormone receptors. For example, scalp, pubic, and beard hair have many androgen receptors and are very sensitive to androgen stimulation —other hair follicles are less responsive.

Kevin McElwee
Wetherby, West Yorkshire

HEATED ARGUMENT

Q **_Why is it that when I get into a bath at 39 °c I feel totally relaxed and, yet, when I enter a room at the same temperature, I feel totally stressed?_**

A Although the bath may be at 39 °c, the air in the bathroom is probably much colder, allowing part of the body to lose excess heat. This heat loss is helped by evaporation of the bath water. In contrast, when the air in a room is 39 °c, heat will be flowing into one's body rather than away, and body temperature will begin to increase. Because the body then has to lose excess heat, the feeling of stress goads the brain into taking action, such as drinking a cold drink or moving to a cooler room.

Keith Lawrence
Staines, Middlesex

INDEX

143-6
45-7
45-6
138-41